Ellie Topp &
Margaret Howard

Put a Lid on It!

Small-Batch Preserving for Every Season

D1318410

Macmillan Canada
Toronto

Canadian Cataloguing in Publication Data

Topp, Ellie
    Put a lid on it! : small batch preserving for every season

Includes index.
ISBN 0-7715-7452-5
1. Canning and preserving.   I. Howard, Margaret.   II. Title.

TX603.T66 1997   641'.4   C96-931975-4

Macmillan Canada wishes to thank the Canada Council, the Ontario Arts Council and the Ontario Ministry of Culture and Communications for supporting its publishing program.

This book is available at special discounts for bulk purchases by your group or organization for sales promotions, premiums, fundraising and seminars. For details, contact: Macmillan Canada, Special Sales Department, 29 Birch Avenue, Toronto, ON M4V 1E2. Tel: 416-963-8830

Cover photo: Hal Roth
Interior photos: Hal Roth (courtesy of Bernardin Ltd.)
Cover design: Gord Robertson

Macmillan Canada
A Division of Canada Publishing Corporation
Toronto, Canada

1 2 3 4 5   KRO   01 00 99 98 97

Printed in Canada

Contents

Authors' Acknowledgements

Put a Lid on It! has enabled us to continue a friendship that began with the writing of *Healthy Home Cooking* (Macmillan 1993). Once again we have thoroughly enjoyed sharing in the research, recipe testing and writing that has brought about this preserving book. In our opinion, such a book, with the latest and most up-to-date information, is long overdue in the Canadian cookbook market. We hope the small batch approach to preserving and the creativity of these recipes will bring as much excitement to your kitchen as it has to ours.

A special recognition is needed for Kevin Myers and Michael Hicks of Hyperactive Productions whose initial enthusiasm for creative preserving inspired both this book and the TV show of the same name on the Life Network.

Once again we have greatly enjoyed working with the staff of Macmillan Canada.

A special thanks to our editor Nicole de Montbrun, copy editor Shaun Oakey and page designer Gordon Robertson who spent many hours transforming our manuscript into a handsome finished book. Also to photographer Hal Roth and associates for their work in creating a most attractive cover.

Thanks are always in order for our supportive and much-loved families and husbands who gave freely of their opinions and of their taste buds. John Howard supplied many hours of editing expertise for the final manuscript and Clarke Topp was instrumental in establishing our means of electronic communication.

We also wish to acknowledge the following people who made significant contributions:

Tom Gleeson of the Health Protection Branch, Health Canada, Ray Beaulieu of the Centre for Food Safety and Applied Nutrition, US Food and Drug Administration, and Richard N. Brown of the United States Department of Agriculture for their help in developing procedures for flavoured oils and vinegars.

Bernardin Canada, Ltd, for use of their photographs and illustrations, and Judi Kingry and Susan Iantorno for technical consultation.

Marian Hebb, who has always helped us in all the legalities, for her attention to the many important details associated with this project.

Marjorie Hollands and Katherine Younker for the analysis and approval of the Canadian Diabetes Association Food Choice Values for our Light 'n' Low Sugar Spreads.

Bonnie Lacroix, Ontario Ministry of Agriculture, Foods and Rural Affairs for reviewing our introductions for accuracy.

Friends and family who so kindly shared their favourite recipe ideas: Norma Billes, Gillian Bolton, Ellen Boynton, Pat Coleman, Sandra Copeland, Marjorie Herron, Janice McDowell, Karen Merseruau, Joyce Pearson, Margaret Slemon, Pat Webber and Fred Yule.

Introduction

M ULTI-HUED peppers, juicy peaches and nectarines, glowing red and purple grapes — all beckon to us at the farmer's market or produce counter. We load our shopping baskets with this bounty from all over the world. And then what? We certainly enjoy eating the fresh produce. But deep within most of us lurks a desire to preserve these flavours for the future. We want to put them in a jar and "put a lid on it"!

Yesterday, our ancestors needed to preserve fresh food to sustain themselves after the short growing season. Today, with a year-round supply of most fruits and vegetables, our reason is more often to transform them — succulent peaches into tantalizing jams for morning toast, mangoes into ketchup to enhance a fillet of fresh fish, juicy tomatoes into snappy salsas or a pickling cucumber into a crisp pickle to add exciting variety to meals. Our book tells you how.

Recipes in *Put a Lid On It!* focus on widely available fruits and vegetables in lots of interesting combinations. You'll notice we offer smaller rather than larger finished amounts. A small yield gives more opportunity to make several different preserves. It also reduces the risk of scorching that is always a danger when cooking larger batches. And it makes large storage areas unnecessary. Most recipes can be made year round and, most important, at your convenience. Our grandmothers spent hours during the growing season preserving a year's supply of food. Today, we have the luxury of preserving what we want, when we choose and in many exotic combinations.

Preserving Techniques

Preserving food at home is not at all difficult. When you decide to "put a lid on it," there are two important things you must do. The first is to destroy all microorganisms such as bacteria, moulds and yeasts naturally present in the food to prevent them from spoiling the preserved food. Having done this, the second thing is to make sure our preserving containers are sealed in such a way that other organisms cannot enter.

Now, a short biology lesson. Microorganisms and enzymes naturally present in foods cause many changes to occur. Not all of these changes are bad. Many microorganisms—bacteria, moulds and yeasts—are intentionally used to create new forms of foods. For instance, bacteria added to milk produce creamy yogurt. Enzymes turn milk into curds, and moulds introduced into the curds create wonderful cheeses. Wine-makers know the result of yeasts growing in grape juice. However, not all organisms cause changes that are desirable; they can cause food to spoil.

Ways to Preserve Food

The four ways we prevent food from spoiling are with heat, acid, sugar and freezing. Two others, salting and drying, do not need a "lid" and so we did not include them in this book.

By Heat

The easiest way to destroy microorganisms present in food is to heat the food. *Processing* is the word traditionally used when filled jars of food are heated to specific temperatures for specific lengths of time. The temperatures and times depend on the density of the food and the size of the jar.

All moulds and yeasts and most bacteria are destroyed at the temperature of boiling water. However, some bacteria, such as *Clostridium botulinum*, can form spores that withstand very high temperatures. Therefore, although these bacteria are destroyed by boiling-water temperatures, their spores may survive. These spores develop into bacteria that are able to grow in an airtight environment (such as a canning jar) and produce a poisonous toxin causing botulism. Fortunately, these bacteria cannot grow in the presence of acids such as vinegar or lemon juice.

For preserving purposes, food is divided into two categories, *high-acid foods* and *low-acid foods*.

High-acid foods are sufficiently acidic to prevent the growth of any spores that can survive boiling-water processing. Most fruits and some tomatoes are high-acid foods. They can be processed at the lower temperatures reached with a boiling-water canner.

Low-acid foods are not sufficiently acidic to inhibit the growth of bacteria spores that can survive boiling-water temperatures. Meats, fish, some tomatoes and all vegetables are low-acid foods. They must be preserved by processing in a pressure canner, which reaches much higher temperatures than can be achieved with boiling-water methods. The canned foods we buy are processed by pressure canning. In this book we don't deal with pressure canning, since few people have the equipment at home.

However, by adding acid, low-acid foods can be made safe for storing after processing at boiling-water temperatures. We use this technique. All our recipes have a pH value falling within the Agriculture Canada/Health Canada definition of high-acid foods. Thus, it is essential to follow accurately the recipes and the recommended canning procedures to preserve food safely.

By Acid

Preserving food using the acid in vinegar is called pickling. If the acid in a food is strong enough, most microorganisms cannot grow. Familiar acids used in this process are many types of vinegars and lemon juice. However, a few microorganisms are able to grow at high acid concentrations. Therefore, it is now recommended that all pickled foods be processed in a boiling-water canner for a short time to destroy all microorganisms.

By Sugar

Sugar in high concentrations traps the water in food, creating an environment where microorganisms cannot grow. Jams and jellies are preserved in this way. Moulds and some yeasts can grow on the surface of such foods, but only in the presence of air. An airtight seal achieved from heat processing prevents the growth of such moulds and yeasts.

By Freezing

Freezing stores food at such low temperatures that no growth can occur. However, some enzyme activity can still go on in vegetables, giving off flavours. To prevent this, vegetables are generally blanched briefly before freezing. Fruits may be frozen in their raw state. Several of our jams, spreads and curds are under a lid and frozen to extend the storage.

When to Preserve Food

Many of our recipes can be made year round with today's wide availability of fresh fruits and vegetables. However, certain foods are available only at certain times of the year. Seville-type oranges, for example, are usually in Canadian stores only during January and February. Other fruits and vegetables, although available throughout the year, may be of better quality at certain times. We believe the quality of our own locally grown produce is superior since it arrives fresh in our kitchens without extended storage. At other times of year, good imported produce is available, but just remember, you may be paying more. So "put a lid on it" when the quality is at its finest and price is at its lowest.

Easy Step-by-Step Preserving

1. Selecting and Preparing Food

The *best* preserves result from using the *best* ingredients. Use produce that is as fresh as possible and at the peak of quality. Most vegetables should be used as soon after picking as possible, but some fruits may require further ripening. Many tender fruits are picked before they are fully ripe, so wait a day or longer until the full flavour has developed. However, most fruits are best for preserving when they are slightly underripe.

Wash the food thoroughly to remove surface dirt and any traces of chemicals. Discard any bruised or mouldy fruit, since microorganisms may have started to grow. Fruit with other surface blemishes or imperfections is fine to use.

2. Preparing Equipment

Partially fill a boiling-water canner with approximately the amount of hot water needed to cover the jars during processing. If food is to be processed less than 10 minutes, sterilize the jars before filling them. Bring the water to a boil and boil the jars for 10 minutes. Even if the jars are not to be sterilized, they need to be kept hot until they are filled. It is helpful to have an extra kettle of boiling water at hand in case the water level needs to be topped up after the filled jars are placed in the canner. If you live in an area with hard water, add a bit of vinegar to the water to prevent a film from forming on the jars.

Follow the manufacturer's directions for treating the metal disc part of the lid. Some are put in hot water for a specified time while others need to be boiled. This process sanitizes the lids and softens the sealing compound so an airtight seal is formed.

3. Filling Canning Jars

The processing time given in our recipes is based on the food being hot when it is put into the jars. It is important that the jars be processed immediately following the cooking stage.

A sterilized funnel is helpful to avoid spills when filling jars. Food may be ladled into the jar or poured using a sterilized small pitcher or measuring cup.

Leave a head space to allow for expansion of food during processing. For most foods, a head space of ½ inch (1 cm) is needed, although the head space may be as little as ¼ inch (5 mm) for sweet spreads. If the jars are too full, the food may boil out and interfere with the formation of the seal. Too much head space may result in the jar not sealing since the processing time is too short to drive out the extra air. We find it easiest to get in the habit of allowing ½ inch (1 cm) for all foods being processed.

Before placing a lid on the jar, be sure to remove any air trapped between pieces of food. Release any air bubbles by sliding a clean small wooden or plastic spatula between glass and food and gently move the food. Then wipe the rim of the jar with a clean cloth to remove any stickiness that could interfere with the formation of the seal.

Remove a metal disc from the hot water and centre it on the jar rim. (Buy a magnetic lid lifter, or glue a small magnet to the end of a wooden dowel rod to make a handy tool for lifting snap lids from the boiling water.) Then apply the screw band *just until it is fingertip tight.* Use only your fingertips! During processing the air in the jar expands and is vented under the lid. When the jar cools, the air contracts and the lid "snaps" down, creating an airtight vacuum seal. If the lid is too tight, the air cannot escape from the jar, possibly resulting in a failed seal.

4. Processing Canning Jars

Place the filled jars on the rack in the boiling-water canner. Adjust the amount of water to cover the jars by at least 1 to 2 inches (2.5 to 5 cm). Cover the canner and return to a boil. Start counting the processing time called for in the recipe when the water has returned to a steady boil. A kitchen timer is helpful for this job. The processing time for each food is based on the size of the jar and the density and composition of the food, so follow times exactly. Under-processing can result in spoiled or off-flavoured food and over-processing may overcook the food.

When the processing is finished, remove the jars from the canner. Use a jar lifter or lift the rack by its handles. Transfer the jars to a wooden cutting board or a surface covered with several layers of dry towels or newspapers. Do not place jars in a draft or on a cold hard surface, or they may break.

Do not tighten the seal; let the jars cool, undisturbed, for 12 to 24 hours. Then check the seal. It is easy to tell if a jar is sealed, as the metal disc curves downwards. Refrigerate any jars that are not sealed for up to three weeks. You may remove the screw bands to prevent rusting during storage. The bands are not necessary, as the firm seal achieved by the canning process is strong enough to keep the jar airtight.

Processing at Higher Altitudes

At higher altitudes water boils at a lower temperature. So it is necessary to increase the processing time if you live at elevations above 1,000 feet (306 m). Adjust your processing times according to the following chart.

Height Above Sea Level	Increase Processing Time
1,001 to 3,000 feet (306 to 915 m)	5 minutes
3,001 to 6,000 feet (916 to 1,830 m)	10 minutes
6,001 to 8,000 feet (1,831 to 2,440 m)	15 minutes
8,001 to 10,000 feet (2,441 to 3,050 m)	20 minutes

5. Storing Preserved Foods

When the jars are cool and you have checked the seals, attach labels with the contents and the date. Preserved foods are best kept in a dark, cool place. Light may cause food to darken and a heat source, such as hot pipes or a furnace or stove, may hasten the loss of quality. A dark cupboard or area in the basement is ideal.

If our recipes and canning procedures are followed carefully, there should be no problem with spoilage. However, when you open a jar of preserved food, it is a good idea to look closely for any sign of spoilage like a bulging lid or any leakage. The lid should be tight and give resistance when opened. If the lid is loose, or if the food has any off odours or mould on the surface, the food should be discarded. Don't take any chances.

Equipment for Safe Boiling–Water Canning

Boiling–Water Canner

A boiling-water canner is a large covered container generally made from steel-covered enamel or stainless steel. A rack fits inside to hold the jars, keeping them from touching one another and elevating them from the bottom of the canner to allow water to circulate freely around them. The canner must be deep enough to allow at least 1 to 2 inches (2.5 to 5 cm) of briskly boiling water to cover the filled jars and the diameter should be no more than 4 inches (10 cm) wider than the burner on the stove.

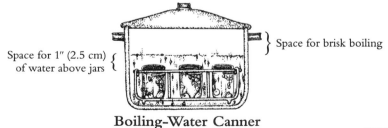

Space for 1″ (2.5 cm)
of water above jars

Space for brisk boiling

Boiling-Water Canner

courtesy of Bernardin

Any large cooking pot can be used for a canner as long as it has a tight-fitting lid and is large enough to hold the jars and allow adequate boiling water to cover the jars. A rack is essential for adequate circulation of water around the jars. A round cake rack can serve this purpose. If the rack does not have handles, you will need a jar lifter to remove the jars from the hot water.

Canning Jars and Lids

Before you start a recipe, be sure you have enough canning jars that are clean and free from cracks or nicks. Canning jars, commonly called mason jars, are designed to withstand the temperatures of boiling-water canning. They are available in a variety of shapes and sizes from small ½-cup (125 mL) to large 2-quart (2 L) jars. Our recipes are designed for small batches and for variety, so we generally use the half-pint (250 mL) and pint (500 mL) sizes. We use the very small jars for gifts.

Jars have two sizes of lids, standard and wide mouth. The standard size is most commonly used, but wide-mouth jars are useful for packing foods in larger pieces, such as dill pickles. Lids are made in two pieces, a metal disc and a screw band to keep the disc in place. The disc has a sealing compound that allows it to form a seal with the jar. Each disc, often called a snap lid, is used one time only to ensure a proper seal. The screw band can be reused.

Possible Causes for Seal Failure or Spoiled Food

- Food was not processed in the canner for the correct time. It's important to start counting processing time just after the water returns to a rapid boil.
- Processing time was not adjusted for altitude (see page 6).
- New sealing discs were not used or were not softened in hot or boiling water.
- Screw bands were put on too tight or were retightened after processing.
- Too much or insufficient head space was left in the jar.
- The jar was cracked before, during or after processing. Cracking during processing could result from adding cold water to a canner of filled jars, placing hot jars on a cold surface or using jars not designed to withstand boiling-water temperatures.
- The amount of vinegar or other acid called for in the recipe was not measured accurately.
- The vinegar was not the standard 5% acetic acid. Always use vinegars of known acidity for canning purposes. For more about vinegar, see page 95.

Sweet
Spreads

Sweet Spreads

M AKING SWEET SPREADS is synonymous with creative home-making. Indeed, the fabulous smells wafting from the kitchen as they cook and the neat rows of jars on the cupboard shelf are both comforting and satisfying evidence of your creativity. But, best of all, the flavour of even the best commercial jams and other spreads does not stand up to the marvellous homemade ones. Furthermore, there are so many fruits that never find their way into a bottled jam. What a pity! Our recipes will encourage you to try exciting combinations never seen on store shelves. And a jar of homemade jam given to a host or hostess is always greatly appreciated. But be sure to include the recipe with it.

Name That Spread!

Jams, preserves, jellies, marmalades, conserves, fruit butters and fruit curds all share the same characteristic consistency from a gel formed by pectin. What makes them different is the size or absence of fruit pieces, the method of cooking and the addition of other ingredients.

Jam is a mixture of fruit and sugar made either from fruits that are high in pectin content or with added pectin. The fruit is usually chopped very finely or mashed.

A *preserve* is the same as a jam but the fruit is in larger pieces.

Jelly is the same as jam except that the cooked fruit has been strained to give a clear spread. Jellies are usually made from fruits high in pectin or with added pectin.

Marmalade is a jam made from citrus fruit. Marmalades generally do not have pectin added since citrus rinds and seeds contain enough pectin to form a soft gel.

A *conserve* is a jam with nuts, dried fruits and often spices added.

Fruit butter is a sweet spread made by cooking fruit pulp with sugar until it has a thick, smooth consistency with no liquid remaining. Spices are often added.

Fruit curd is a sweet spread made from citrus fruit, sugar, butter and eggs cooked gently until thickened.

Essential Sweet Spread Ingredients

The essential ingredients for sweet spreads are fruit, sugar, acid (in the form of citrus juice) and sometimes added pectin. Pectin, whether naturally present in fruit or added, is the substance that causes a fruit to form a gel. But it is not enough for these ingredients to simply be present — the proportion between them is critical to forming a gel. Following tested recipes, such as the ones found in this book or those supplied by pectin manufacturers, will give you the best chance of success.

For more flavour, we like to add such extras as small amounts of liqueurs, nuts, spices and citrus zest. Generally, these are added at the last minute, just before bottling.

Fruit

Fruit should be firm and ripe or, ideally, just underripe, and always of good quality. Never use overripe fruit, since the pectin decreases as the fruit ripens, resulting in a jam that may not form a gel. Slightly underripe fresh fruits contain the most pectin. Irregular-shaped fruit or fruit that is scarred is perfectly good, but discard any that is spoiled or mouldy. Always wash or rinse fruit before use to remove any traces of dust, dirt or chemicals with which it may have been treated.

Fruit frozen without sugar is great for making jam. Plan to freeze such fruits as rhubarb, berries and cherries when they are plentiful to make into jam later at your convenience. Choose clean, slightly underripe fruits at the peak of the growing season. Place the fruit in single layers on shallow cookie trays in the quantities required for each recipe you plan to use. Freeze the trays of fruit and then package in airtight labelled containers. You don't need to defrost the fruit before using it in the recipe.

Sugar

Sugar is a vital ingredient in jams and jellies. Its combination with pectin is essential to the gelling process. It influences taste by enhancing the natural flavour of the fruit. High concentrations of sugar play a major role in preservation by preventing the growth of microorganisms.

Sugar affects the texture of the fruit used in sweet spreads. If fruit and sugar are simply cooked together, the fruit quickly breaks down. However, when sugar is combined with fruit for several hours before cooking, the fruit shrinks as the sugar draws out some of its juice. This partially dehydrated fruit keeps its shape in the finished spread. Our elegant Sherried Strawberry Preserves (page 20) is a perfect example.

Generally, granulated sugar is preferred, but occasionally brown is used. Avoid very dark brown sugar, as it will change the colour of the spread. Keeping quality is not affected by the type of sugar used. For our Light 'n' Low-Sugar Spreads, a combination of some sugar with an artificial sweetener is often used.

Pectin and Gel Formation

Pectin, a natural gumlike substance, is found in many fruits and vegetables. Different fruits contain pectin in varying amounts, with the highest concentration found in the cores, seeds and skins. It is pectin that gives sweet spreads their smooth, semisolid consistency. Some fruits have enough pectin to make products set well, while others require added pectin.

As well as pectin, fruit must contain the correct amount of acid to form a gel. Too much acid results in a gel that sets too quickly and too firmly, making the sweet spread "weep" as moisture is squeezed out. Marmalades often have too much acid, so baking soda must be added to reduce their acidity. Too little acid prevents a gel from forming. Lemon juice is added to low-acid fruits to increase their acidity.

It's the balance between fruit, sugar and acid that is critical for gel formation. This balance is a delicate one, so it is essential that you measure all ingredients accurately. Do not change their amounts, especially the amounts of sugar and acid. Even then, a gel is not guaranteed. The natural sugar, pectin and acid content of fruit always varies slightly because of weather and storage conditions. Some types of gels, especially in marmalades, may require several hours or even days for the pectin to set. Others form quickly, even before completely cooling.

When making sweet spreads from only one kind of low-pectin fruit, you need to add pectin. Increasing the pectin level produces a fresher-tasting jam by decreasing the otherwise long cooking time necessary to concentrate the natural pectin. Adding more pectin requires adding more sugar (to retain the correct balance), resulting in a greater amount of sweet spread for the amount of fruit used.

There are three ways to add pectin to fruit.

- **Commercial pectin:** Be sure to check the "best before" date and follow the manufacturer's instructions.

- **Homemade apple pectin:** Make your own pectin from apples (page 35). This is an interesting source that can be economical if you have an abundance of apples.

- **High-pectin/low-pectin fruits:** Combine high-pectin fruits with low-pectin ones. Adding apple to rhubarb, or red currants to raspberries, are good examples.

Guide to Pectin Content of Fruits

Fruits marked with an asterisk (★) have sufficient acid to combine with pectin for gel formation.

High-Pectin Fruits	Low-Pectin Fruits
apples (sour★ and sweet)	apricots★
cherries (sour★ and sweet)	blueberries
crabapples★	elderberries
cranberries★	figs
currants (red★ and black★)	nectarines
gooseberries★	peaches
grapefruit★	pears
grapes★	pineapple
kiwifruit★	raspberries★
lemons★, limes★	rhubarb★
oranges★	strawberries★
quinces	
plums (some kinds)★	

Tests for Determining When a Gel Will Form

There are two tests to determine when a sweet spread will form a gel. Marmalades and conserves form a lighter gel than jams and jellies.

Sheet or Spoon Test

Begin cooking. To test for gel formation, periodically dip a cool metal spoon into the hot fruit mixture and immediately lift the spoon so the mixture runs off the side. At first the drops will be light and syrupy. As the mixture continues to cook, the drops from the spoon will become heavier. When the mixture "sheets" from the spoon (the drops become very thick and two drops run together before dropping off), it will form a gel on cooling.

Freezer Test

Place two small plates in the freezer ahead of time. To test for gel formation, put a spoonful of hot fruit mixture on one chilled plate. Immediately return it to the freezer and wait for 2 minutes. Meanwhile, remove the saucepan from the heat to prevent overcooking. If the mixture is sufficiently cooked, it will form a gel that moves slowly as the plate is tilted. If it doesn't form a gel, cook for another two minutes and repeat the test on the remaining chilled plate.

Processing Sweet Spreads

Sweet spreads are preserved by freezing or by 5-minute processing in a boiling water canner. Detailed instructions for boiling-water processing can be found on page 4. Since the processing time for sweet spreads is less than 10 minutes, it is essential to sterilize the jars and any equipment used for filling them. Jars with two-piece lids must be used to develop an airtight seal. If not processed, sweet spreads may be stored in the refrigerator for up to three weeks.

For many years paraffin has been used to seal jam jars, but it is no longer recommended. A layer of paraffin on top of a sweet spread does not give the necessary airtight seal. Moulds may grow in the small cracks and pinholes that occur when the wax cools. We used to think that simply removing the mould was sufficient to make the jam safe. However, research has since found that mould growth may produce harmful substances that can penetrate unseen throughout the jar. So invest in some of the attractive small canning jars that may be processed in a boiling-water bath. They are safe as well as pretty!

Essential Sweet Spread Equipment

A *large saucepan* is essential to allow the fruit mixture to come to a full rolling boil. It should be heavy to allow even distribution of heat and made of stainless steel or enamel to prevent reaction with the acid in the mixture. In our recipes, a "large saucepan" means one that holds approximately 4 quarts (4 L). A few of the larger recipes call for a "very large saucepan," meaning one that holds at least 6 quarts (6 L).

Sweet spreads should be preserved in mason jars closed with a screw band and a new metal disc. It is best to use half-cup (125 mL) or half-pint (250 mL) mason jars. Larger jars may result in a spread with a softer gel since the longer cooling time of the larger quantity can cause breakdown of the gel.

You will also need a *boiling-water canner* with a basket for processing the fruit mixture, a *ladle or pitcher* for putting the fruit mixture into jars, and *tongs or a jar lifter* for lifting the jars from the boiling water (unless the canner rack has handles).

Procedure for Shorter Time Boiling–Water Processor

On the following page is the step-by-step procedure for the processing of foods that require less than 10 minutes processing time. Use this procedure for all sweet spreads as directed in the recipes.

Shorter Time Processing Procedure

(For food that requires less than 10 minutes processing time.)

If the recipe requires a preparation and cooking time longer than 20 minutes, begin preparation of the ingredients first. Then bring the water and jars in the canner to a boil while the prepared food is cooking. If the ingredients require a shorter preparation and cooking time, begin heating the canner before you start your recipe. Remember the jars need to be sterilized before you fill them, which requires 10 minutes after the water reaches a boil. Have a kettle with boiling water handy to top up the water level in the canner after you have put in the jars.

Steps for Perfect Processing

20 Minutes Before Processing
Partially fill a boiling water canner with hot water. Place the number of clean mason jars needed to hold the quantity of finished food prepared in the recipe into the canner. Cover and bring the water to a boil over high heat. Boil for at least 10 minutes to sterilize jars. This step generally requires 20 to 30 minutes, depending on the size of your canner.

5 Minutes Before Processing
Approximately 5 minutes before you are ready to fill the jars, place snap lids in hot or boiling water according to manufacturer's directions.

Filling Jars
Remove jars from canner and pour or ladle the foods into hot jars to within ½ inch (1 cm) of top rim (head space). If the food is in large pieces, remove trapped air bubbles by sliding a rubber spatula between glass and food; readjust the head space to ½ inch (1 cm). Wipe jar rim to remove any stickiness. Centre snap lid on jar; apply screw band just until fingertip tight.

Processing Jars
1. Place jars in canner and adjust water level to cover jars by 1 to 2 inches (2.5 to 5 cm). Cover canner and return water to boil. Begin timing when water returns to a boil. Process for 5 minutes.
2. Remove jars from canner to a surface covered with newspapers or with several layers of paper towels and cool for 24 hours. Check jar seals (sealed lids turn downward). Label jars with contents and date and store in a cool, dark place.

Chapter One

Jams for All Seasons

C ERTAINLY the expression "a jam for all seasons" is right at home in this book. We offer you jams for "in season" to make when our superb homegrown produce is available. Other jams can be made almost any time of year with the great availability of imported produce, albeit at a price. Quick-to-make freezer and microwave jams are also included. Our emphasis has been on variety, so the majority of the recipes are for small batches.

Be creative with jams. Homemade Apple Pectin (page 35), made when apples are plentiful, can be used with any quantity of fruit. This allows you to make as little as one jar of jam. Thus, small quantities of leftover fruit can be readily converted into interesting jam. For example, one day we had a couple of pears approaching ripeness. Adding a few frozen blueberries along with apple pectin, sugar and lemon juice made one jar of a great-tasting jam. With apple pectin handy on the shelf, preparing the jam took less than 20 minutes.

Some fruits you may want to use for jam do not contain enough pectin to form a gel in a reasonable cooking time. To get a nice gel in a short cooking time, add commercial pectin, our Homemade Apple Pectin or fruits with a higher pectin content.

We have discovered that adding warmed sugar to our uncooked freezer jams helps the sugar to dissolve in the fruit. This process is also useful in Old-fashioned Raspberry Jam (page 21), as raspberries are low in pectin. Warmed sugar dissolves faster in the simmering fruit, thereby protecting the pectin from breaking down.

Serving Suggestions

Imagine a small tart filled with a flavourful fruit jam. Or that same jam as a filling between cake layers. We like making our own fruit-flavoured yogurts by adding a spoonful of jam to plain yogurt.

List of Recipes

Favourite Strawberry Jam

Generations have made strawberry jam to preserve this favourite summer fruit. Traditionally, low-pectin strawberries are cooked for long periods to achieve a gel. Our method uses standing periods alternating with much shorter cooking times. It makes a jam that retains its lovely red colour and fresh flavour.

4 cups	halved or quartered firm strawberries, depending on size	1 L
2 cups	granulated sugar	500 mL
¼ cup	lemon juice	50 mL

1. Mix berries and sugar and let stand for 8 hours, stirring occasionally.
2. Place berry mixture in a medium stainless steel or enamel saucepan. Bring to a boil over medium heat. Add lemon juice, return to a boil and boil rapidly for 5 minutes. Remove from heat, cover and let stand for 24 hours.
3. Bring berries to a full boil over high heat and boil rapidly for 5 minutes, stirring constantly. Remove from heat.
4. Ladle into sterilized jars and process as directed on page 16 (Shorter Time Processing Procedure).

Makes 2½ cups (625 mL).

Variation:

Strawberry Rhubarb Jam: Add 1 cup (250 mL) finely chopped rhubarb to strawberries in step 1. Makes 3 cups (750 mL).

Sherried Strawberry Preserve

Whole strawberries with a hint of sherry are suspended in this delightful preserve. A perfect accompaniment to fresh biscuits. Stir before serving to break gel and distribute fruit.

5 cups	whole small firm strawberries (about 2½ pints)	1.25 L
4 cups	granulated sugar	1 L
3 tbsp	lemon juice	45 mL
1	pouch (85 mL) liquid fruit pectin	1
½ cup	medium-dry sherry	125 mL

1. Stir together berries, sugar and lemon juice in a large bowl. Cover and let stand for 4 hours, stirring occasionally.
2. Place berries in a very large stainless steel or enamel saucepan. Bring to a boil over high heat and boil rapidly for 2 minutes, stirring constantly. Remove from heat; stir in pectin and sherry.
3. Ladle into sterilized jars and process as directed on page 16 (Shorter Time Processing Procedure).

Makes 5½ cups (1.4 L).

Variation:

Strawberry Preserves with White Wine: Use white wine in place of the sherry for a more delicate flavour.

Old-Fashioned Raspberry Jam

The intense raspberry flavour of this jam makes it a long-time favourite. Warming the sugar beforehand keeps the jam boiling evenly and ensures success.

| 4 cups | granulated sugar | 1 L |
| 4 cups | raspberries | 1 L |

1. Place sugar in an ovenproof shallow pan and warm in a 250°F (120°C) oven for 15 minutes. (Warm sugar dissolves better.)
2. Place berries in a large stainless steel or enamel saucepan. Bring to a full boil over high heat, mashing berries with a potato masher as they heat. Boil hard for 1 minute, stirring constantly.
3. Add warm sugar, return to a boil and boil until mixture will form a gel,★ about 5 minutes.
4. Ladle into sterilized jars and process as directed on page 16 (Shorter Time Processing Procedure).

Makes 4 cups (1 L).

Tip: To make a small boiling-water canner, tie several screw bands together with string or use a small round cake rack in the bottom of a large covered Dutch oven. Be sure the pan is high enough for 2 inches (5 cm) of water to cover the jars when they are sitting on the rack.

★To determine when mixture will form a gel, see page 14.

Raspberry Jam with Chambord

This elegant jam is an ideal gift for a special friend. The unparalleled flavour of fresh raspberries is wonderfully complemented by raspberry liqueur.

3¾ cups	crushed raspberries	925 mL
	(about 5 cups/1.25 L whole berries)	
4 cups	granulated sugar	1 L
3 tbsp	lemon juice	45 mL
1	pouch (85 mL) liquid fruit pectin	1
⅓ cup	Chambord, framboise or	75 mL
	Raspberry Liqueur (page 208)	

1. Combine berries, sugar and lemon juice in a very large stainless steel or enamel saucepan. Let stand for 10 minutes.
2. Place fruit over high heat, bring to a full boil and boil hard for 2 minutes, stirring constantly. Remove from heat; stir in pectin and liqueur.
3. Ladle into sterilized jars and process as directed on page 16 (Shorter Time Processing Procedure).

Makes 6 cups (1.5 L).

Tip: To add amazing flavour to a simple cheesecake, swirl ⅔ cup (150 mL) Raspberry Jam with Chambord through the cake before baking.

Red and Black Currant Cassis Jam

The strong, rich taste of two kinds of currants reinforced by a currant-based liqueur makes this a jam for the true currant lover. Currants are very high in pectin, so don't overcook them. They thicken up considerably after cooking.

2½ cups	black currants, washed and stemmed	625 mL
2½ cups	red currants, washed and stemmed	625 mL
1 cup	water	250 mL
3 cups	granulated sugar	750 mL
2 tbsp	lemon juice	25 mL
3 tbsp	Cassis	45 mL

1. Place fruit and water in a large stainless steel or enamel saucepan. Bring to a boil over high heat, cover, reduce heat and boil gently for 10 minutes, stirring occasionally.
2. Add sugar, lemon juice and liqueur. Bring to a full boil over high heat, stirring constantly. Boil rapidly uncovered until mixture will form a gel,★ about 10 minutes. Remove from heat.
3. Ladle into sterilized jars and process as directed on page 16 (Shorter Time Processing Procedure).

Makes 5 cups (1.25 L).

★To determine when mixture will form a gel, see page 14.

Spiced Blueberry Honey Jam

Honey adds its own delicate nuance to the more defined blueberry and nutmeg flavours in this delightful jam. It can be made any time of year with frozen blueberries.

2½ cups	fresh or frozen coarsely chopped blueberries	625 mL
2½ cups	granulated sugar	625 mL
1 cup	liquid honey	250 mL
1 tbsp	lemon juice	15 mL
½ tsp	ground nutmeg	2 mL
1	pouch (85 mL) liquid fruit pectin	1

1. Place blueberries, sugar, honey, lemon juice and nutmeg in a large stainless steel or enamel saucepan. Bring to a full boil over high heat and boil hard for 2 minutes, stirring constantly. Remove from heat and stir in pectin.
2. Ladle into sterilized jars and process as directed on page 16 (Shorter Time Processing Procedure).

Makes 4 cups (1 L).

Tip: Use a small grater to grate the seed of a nutmeg for freshest flavour.

Bluebarb Jam

This combination of blueberry and rhubarb has become one of our favourite jams. Only by using frozen fruit can we make this wonderful jam from fruits having different growing seasons. Those of you living in the prairies can substitute saskatoonberries for blueberries.

3½ cups	chopped fresh or frozen rhubarb	875 mL
½ cup	water	125 mL
2¼ cups	coarsely chopped fresh or frozen blueberries	550 mL
1 tbsp	lemon juice	15 mL
1	box (57 g) pectin crystals	1
5½ cups	granulated sugar	1.375 L

1. Place rhubarb and water in a large stainless steel or enamel saucepan. Bring to a boil over high heat, cover, reduce heat, and simmer for 5 minutes, stirring frequently.
2. Add blueberries, lemon juice and pectin; mix well. Bring to boil over high heat, stirring constantly. Add sugar, return to a full boil and boil hard for 1 minute, stirring constantly. Remove from heat.
3. Ladle into sterilized jars and process as directed on page 16 (Shorter Time Processing Procedure).

Makes 6 cups (1.5 L).

Gooseberry Rhubarb Jam

This combination of two tart, old-fashioned country garden fruits gives us a jam with a wonderful flavour and glorious colour. If you are making this jam with frozen fruit, chop and measure while fruit is still frozen.

2 cups	finely chopped rhubarb	500 mL
½ cup	water	125 mL
2 cups	gooseberries, stems removed and coarsely chopped	500 mL
2 tbsp	lemon juice	30 mL
5½ cups	granulated sugar	1.375 L
1	pouch (85 mL) liquid fruit pectin	1

1. Place rhubarb and water in a large stainless steel or enamel saucepan. Bring to a boil over high heat, reduce heat, cover and boil gently for 3 minutes.
2. Stir gooseberries, lemon juice and sugar into rhubarb. Return to a full boil over high heat and boil hard for 1 minute, stirring constantly. Remove from heat and stir in pectin.
3. Ladle into sterilized jars and process as directed on page 16 (Shorter Time Processing Procedure).

Makes 5 cups (1.25 L).

Variation:

Gingered Gooseberry Rhubarb Jam: Add ⅓ (75 mL) finely chopped crystallized ginger in step 2.

Plum and Crabapple Jam

Crabapples are more commonly used in jellies than jams. Combined with plums they impart a sweet-tart flavour and a gorgeous colour to this quite different jam. Do not overcook it. Since plums and crabapples are naturally high in pectin, this jam thickens considerably after it's cooked.

3 cups	quartered unpeeled crabapples (about 4 cups/1 L whole fruit)	750 mL
1½ cups	water	375 mL
1	cinnamon stick	1
4 cups	sliced blue or purple plums (about 8 large or 16 small)	1 L
5 cups	granulated sugar	1.25 L
¾ cup	dry red or white wine or grape juice	175 mL

1. Place crabapples, water and cinnamon stick in a large stainless steel or enamel saucepan. Bring to a boil over high heat, cover, reduce heat and boil gently for 10 minutes or until fruit is soft. Remove from heat and discard cinnamon stick. Press crabapples through a sieve; discard solids.
2. Return crabapple pulp to saucepan. Add plums, sugar and wine. Bring to a full boil and boil rapidly, uncovered, until mixture will form a gel,★ about 20 minutes, stirring frequently. Remove from heat.
3. Ladle into sterilized jars and process as directed on page 16 (Shorter Time Processing Procedure).

Makes 6 cups (1.5 L).

★To determine when mixture will form a gel, see page 14.

Plum Amaretto Jam

The almond flavour of the liqueur nicely complements the tartness of the plums in this rich purple jam. Plums are seldom used for jam, but after tasting this one you'll want to take a plum to breakfast more often.

3 cups	chopped tart red or purple plums (about 8 to 10 plums)	750 mL
3 cups	granulated sugar	750 mL
¼ cup	water	50 mL
3 tbsp	lemon juice	45 mL
¼ cup	Amaretto	50 mL

1. Combine plums, sugar, water and lemon juice in a large stainless steel or enamel saucepan. Bring to a boil over high heat, reduce heat to medium and boil rapidly until mixture will form a gel,★ about 20 minutes, stirring frequently. Remove from heat and stir in liqueur.
2. Ladle into sterilized jars and process as directed on page 16 (Shorter Time Processing Procedure).

Makes 4 cups (1 L).

★To determine when mixture will form a gel, see page 14.

Peach Pear Jam with Lime

Two fall fruits combine to make one of our favourite jams. This started off as Ellie's peach recipe. One day, with not enough peaches to make the jam, she added some pears. Since these fruits are best in season for making this jam, this is a larger recipe than many others.

	rind of 1 lime	
2 cups	finely chopped peeled peaches	500 mL
2 cups	finely chopped peeled pears	500 mL
1	box (57 g) fruit pectin crystals	1
5 cups	granulated sugar	1.25 L

1. Remove thin outer rind from lime with vegetable peeler and cut into fine strips with scissors or sharp knife; or use a zester. Place lime rind in a small microwavable container with ¼ cup (50 mL) water. Microwave on High (100%) for 1 minute. Drain and discard liquid; reserve rind.
2. Place peaches, pears, lime rind and pectin in a very large stainless steel or enamel saucepan. Bring to a boil over high heat, stirring constantly. Add sugar, return to a full boil and boil hard for 1 minute, stirring constantly. Remove from heat.
3. Ladle into sterilized jars and process as directed on page 16 (Shorter Time Processing Procedure).

Makes 7 cups (1.75 L).

Autumn Fruit Jam

These three fruits are all in season at the same time. Together they make a jam that reflects the luscious essence of early-fall fruits. The high pectin content of plums and apples compensates for the low pectin in pears to produce a well-set jam.

5	plums, sliced	5
2	medium apples, peeled, cored and chopped	2
2	medium pears, peeled, cored and chopped	2
1 cup	water	250 mL
2 tsp	grated lemon rind	10 mL
2 tbsp	lemon juice	25 mL
3 cups	granulated sugar	750 mL
½ tsp	each cinnamon and ground ginger	2 mL

1. Combine plums, apples, pears, water, lemon rind and lemon juice in a medium stainless steel or enamel saucepan. Bring to a boil over high heat, cover, reduce heat and cook for 10 minutes or until fruit is softened.
2. Add sugar to fruit and return to a boil, stirring constantly until sugar is dissolved. Boil rapidly, uncovered, until mixture will form a gel,★ about 30 minutes, stirring occasionally. Stir in cinnamon and ginger.
3. Ladle into sterilized jars and process as directed on page 16 (Shorter Time Processing Procedure).

Makes 4 cups (1 L).

Variations:

Replace cinnamon and ginger with 1 tbsp (15 mL) vanilla extract added to cooked jam just before bottling.

Nectarine Plum Apple Jam: Use 4 nectarines, peeled and chopped, instead of pears.

★To determine when mixture will form a gel, see page 14.

Papaya and Pineapple Freezer Jam

Uncooked, the fresh flavours of papaya, pineapple and lime bring the taste of the tropics to this pale golden succulent jam. We have used this jam as a dessert topping or folded into plain yogurt as a simple dessert.

3½ cups	granulated sugar	875 mL
1	lime	1
1	papaya, peeled, seeded and finely chopped	1
¾ cup	finely chopped fresh pineapple, peeled and cored	175 mL
1	pouch (85 mL) liquid fruit pectin	1

1. Place sugar in an ovenproof shallow pan and warm in a 250°F (120°C) oven for 15 minutes. (Warm sugar dissolves better.)
2. Remove thin outer rind from lime with vegetable peeler and cut into fine strips with scissors or sharp knife; or use a zester. Place lime rind in a small microwavable container with ¼ cup (50 mL) water. Microwave on High (100%) for 1 minute. Drain and discard liquid; reserve rind. Squeeze lime and reserve juice.
3. Place papaya in a 2-cup (500 mL) measuring cup and add enough chopped pineapple to make 1¾ cups (425 mL) fruit. Transfer to a large bowl. Stir in lime rind and warmed sugar and let stand for 10 minutes, stirring occasionally.
4. Stir in 2 tbsp (25 mL) lime juice and pectin, stirring constantly for 3 minutes.
5. Ladle jam into clean jars or plastic containers to within ½ inch (1 cm) of rim. Cover with tight-fitting lids. Label jars and let stand at room temperature until set, up to 24 hours.
6. Refrigerate for up to 3 weeks or freeze for longer storage.

Makes 4 cups (1 L).

Blueberry Freezer Jam
with Cointreau

Orange highlights the intense blueberry taste of this freezer jam. Freezing the jam eliminates cooking and retains the fresh flavour of the fruit.

2½ cups	granulated sugar	625 mL
1½ cups	crushed blueberries	375 mL
	(about 2 cups/500 mL whole fruit)	
1	orange, peeled and finely chopped	1
1 tbsp	Cointreau or other orange liqueur	15 mL
1	pouch (85 mL) liquid fruit pectin	1

1. Place sugar in an ovenproof shallow pan and warm in a 250°F (i20°C) oven for 15 minutes. (Warm sugar dissolves better.)
2. Combine blueberries, orange and sugar in a large bowl and let stand for 10 minutes, stirring occasionally.
3. Stir in liqueur and pectin, stirring constantly for 3 minutes.
4. Ladle jam into clean jars or plastic containers to within ½ inch (1 cm) of rim. Cover with tight-fitting lids. Label jars and let stand at room temperature until set, up to 24 hours.
5. Refrigerate for up to 3 weeks or freeze for longer storage.

Makes 3 cups (750 mL).

Cherry Orange Freezer Jam

What a treat to have this fresh-tasting jam during the winter prepared from sour cherries you froze last summer. And if you didn't freeze any, you can find them frozen in bulk food stores in the dead of winter. It helps if the cherries have been pitted before freezing.

2½ cups	granulated sugar	625 mL
1	medium orange	1
1½ cups	chopped fresh or frozen pitted sour cherries	375 mL
¾ cup	water	175 mL
1	box (57 g) fruit pectin crystals	1

1. Place sugar in an ovenproof shallow pan and warm in a 250°F (120°C) oven for 15 minutes. (Warm sugar dissolves better.)
2. Meanwhile, remove thin outer rind from orange with vegetable peeler and cut into fine strips with scissors or sharp knife; or use a zester. Remove and discard remaining white rind. Finely chop orange pulp with a knife or in a food processor with on/off motion.
3. Place rind and pulp in large bowl. Stir in cherries and warm sugar. Let stand for 10 minutes, stirring occasionally.
4. Combine water with pectin in a small saucepan. Boil for 1 minute, stirring constantly. Stir pectin mixture into fruit, stirring constantly for 3 minutes.
5. Ladle jam into clean jars or plastic containers to within ½ inch (1 cm) of rim. Cover with tight-fitting lids. Label jars and let stand at room temperature until set, up to 24 hours.
6. Refrigerate for up to 3 weeks or freeze for longer storage.

Makes 4 cups (1 L).

Minted Raspberry Peach Freezer Jam

The freshness of mint enhances this uncooked classic fruit combination.

3½ cups	granulated sugar	875 mL
1 cup	mashed fresh or frozen unsweetened raspberries	250 mL
⅔ cup	finely chopped peeled peaches (1 large peach)	150 mL
2 tbsp	finely chopped fresh mint	25 mL
2 tbsp	lemon juice	25 mL
1	pouch (85 mL) liquid fruit pectin	1

1. Place sugar in an ovenproof shallow pan and warm in a 250°F (120°C) oven for 15 minutes. (Warm sugar dissolves better.)
2. Combine raspberries, peaches, mint and warm sugar in a large bowl and let stand for 10 minutes, stirring occasionally.
3. Add lemon juice and pectin, stirring constantly for 3 minutes.
4. Ladle jam into clean jars or plastic containers to within ½ inch (1 cm) of rim. Cover with tight-fitting lids. Label jars and let stand at room temperature until set, up to 24 hours.
5. Refrigerate for up to 3 weeks or freeze for longer storage.

Makes 4 cups (1 L).

Homemade Apple Pectin

For the adventurous purist or the economy-minded with an ample supply of apples, pectin making is an interesting and satisfying project. Apple pectin imparts a delicate apple flavour to any jam made from it. Making your own pectin lets you create small batches of unusual jams. Two combinations we like are pears with blueberries and mango with kiwifruit. The pectin content of apples decreases during storage, so make Homemade Apple Pectin in the fall when apples are at their freshest.

7	tart apples (about 2 lb/1 kg)	7
4 cups	water	1 L
2 tbsp	lemon juice	25 mL

1. Cut apples into quarters (do not peel or core). Combine with water and lemon juice in a large stainless steel or enamel saucepan. Bring to a boil over high heat, cover, reduce heat and simmer for 40 minutes, stirring occasionally.
2. Strain through a coarse sieve and discard solids. Pour liquid through a jelly bag or several layers of cheesecloth.
3. Ladle into sterilized jars and process as directed on page 16 (Shorter Time Processing Procedure).

Makes 4 cups (1 L).

To Use Homemade Apple Pectin for Jam or Jelly
1. For each 1 cup (250 mL) of finely chopped fruit or juice, use 1 cup (250 mL) Homemade Apple Pectin and ¾ cup (175 mL) granulated sugar.
2. Combine in a stainless steel or enamel saucepan with 1 tsp (5 mL) lemon juice if fruit is low acid (see chart on page 13). Bring to a boil over high heat and boil rapidly, stirring constantly, until mixture will form a gel,★ about 10 minutes.
3. Ladle into sterilized jars and process as directed on page 16 (Shorter Time Processing Procedure).

★To determine when mixture will form a gel, see page 14.

Tropical Pineapple Winter Jam

This is the jam to make on a stormy day in January when our great Canadian fruits are in short supply. For a moment, the taste and smell of citrus fruits, pineapple and coconut will transport you to the tropics.

1	each large orange, lime, lemon	1
1	can (19 oz/540 mL) crushed pineapple packed in juice	1
2	medium tart apples, peeled, cored and finely chopped	2
1½ cups	granulated sugar	375 mL
½ cup	sweetened coconut milk	125 mL
2 tbsp	rum OR 1 tsp (5 mL) rum extract	25 mL
¼ tsp	ground nutmeg	1 mL

1. Remove thin outer rind from orange, lime and lemon with vegetable peeler and cut into fine strips with scissors or sharp knife; or use a zester. Remove and discard remaining white rind. Finely chop orange, lime and lemon pulp with a knife or in food processor with on/off motion.
2. Place rind and chopped pulp in a large stainless steel or enamel saucepan. Drain pineapple, reserving ¼ cup (50 mL) juice. Add pineapple, reserved juice, apples, sugar and coconut milk to saucepan. Bring to a full boil over high heat, reduce heat and boil gently for 20 minutes, stirring occasionally.
3. Add rum and nutmeg to jam. Continue to boil until mixture will form a gel,★ about 5 minutes.
4. Ladle into sterilized jars and process as directed on page 16 (Shorter Time Processing Procedure).

Makes 4 cups (1 L).

Tip: Coconut milk can be made by processing 3 cups (750 mL) sweetened coconut and 1⅓ (325 mL) hot water in a blender for 10 seconds. Strain through a coffee filter or cheesecloth before using.

★To determine when mixture will form a gel, see page 14.

Raspberry and Blueberry Jam

The flavours of the two berries and the citrus fruit combine beautifully in this interesting jam. Make this jam year round from either fresh or frozen berries.

3 cups	fresh or frozen unsweetened raspberries	750 mL
2 cups	fresh or frozen unsweetened blueberries	500 mL
1	large orange	1
6½ cups	granulated sugar	1.625 L
2 tbsp	lemon juice	25 mL
1	pouch (85 mL) liquid fruit pectin	1

1. Mash raspberries and blueberries in a large stainless steel or enamel saucepan.
2. Remove thin outer rind from orange with vegetable peeler and cut into fine strips with scissors or sharp knife; or use a zester. Add to saucepan. Remove and discard remaining white rind. Finely chop orange in food processor with on/off motion to measure ½ cup (125 mL). Add orange pulp, sugar and lemon juice to saucepan.
3. Bring fruit to a full boil over high heat and boil hard for 1 minute, stirring constantly. Remove from heat and stir in pectin.
4. Ladle into sterilized jars and process as directed on page 16 (Shorter Time Processing Procedure).

Makes 7 cups (1.75 L).

Variations:

Raspberry Cranberry Jam: Replace blueberries with 2 cups (500 mL) fresh or frozen cranberries, finely chopped.
Raspberry Plum Jam: Replace blueberries with 2 cups (500 mL) finely chopped plums. A small amount of finely chopped fresh mint makes a nice addition.

Cranberry Pear Lemon Jam

Combine these three distinctly flavoured fruits for a delicious breakfast jam. Its tart tangy flavour also marries well with roasted poultry and pork.

4	large Bartlett pears, peeled, cored and diced (about 4 cups/1 L)	4
3 cups	coarsely chopped fresh or frozen cranberries	750 mL
½ cup	water	125 mL
2 tsp	grated lemon rind	10 mL
2 tbsp	lemon juice	25 mL
1¾ cups	granulated sugar	425 mL

1. Combine pears, cranberries, water and lemon rind in a large stainless steel or enamel saucepan. Bring to a boil over high heat, cover, reduce heat and cook for 5 minutes, stirring frequently.
2. Gradually add lemon juice and sugar, stirring until sugar is dissolved. Boil rapidly, uncovered, until mixture will form a gel,★ about 15 minutes, stirring frequently. Remove from heat.
3. Ladle into sterilized jars and process as directed on page 16 (Shorter Time Processing Procedure).

Makes 5 cups (1.25 L).

★To determine when mixture will form a gel, see page 14.

Winter Pear and Apricot Jam

Bosc and Anjou pears keep for long periods, making them available for most of the winter. They are firmer than other varieties such as the Bartlett and Pakham. The tartness of dried apricots combines with the sweetness of pears to make a great winter jam.

½ cup	finely chopped dried apricots	125 mL
1¾ cups	water	425 mL
2 cups	finely chopped cored, peeled winter pears (Bosc or Anjou)	500 mL
2 tsp	lemon juice	10 mL
1	box (57 g) fruit pectin crystals	1
4½ cups	granulated sugar	1.125 L

1. Soak apricots in water for 4 hours or overnight.
2. Pour apricots into a very large stainless steel or enamel saucepan. Add pears, lemon juice and pectin. Bring to a full boil over high heat, stirring constantly. Add sugar, return to a full boil and boil hard for 1 minute, stirring constantly. Remove from heat.
3. Ladle into sterilized jars and process as directed on page 16 (Shorter Time Processing Procedure). Note that this jam is slow to set.

Makes 6 cups (1.5 L).

Festive Kiwifruit Daiquiri Jam

This is a favourite of people who like jams that are less sweet. It's a beautiful emerald green.

1½ cups	finely chopped kiwifruit (about 5 kiwifruit)	375 mL
⅓ cup	lime juice (about 2 limes)	75 mL
¼ cup	water	50 mL
2 tbsp	dried cranberries or cherries, coarsely chopped	25 mL
3 cups	granulated sugar	750 mL
1	pouch (85 mL) liquid fruit pectin	1
¼ cup	dark rum	50 mL

1. Place kiwifruit, lime juice, water, cranberries and sugar in a large stainless steel or enamel saucepan. Bring to a full boil over high heat and boil hard for 2 minutes, stirring constantly. Remove from heat; stir in pectin and rum.
2. Ladle into sterilized jars and process as directed on page 16 (Shorter Time Processing Procedure).

Makes 4 cups (1 L).

Mango Plum Jam

A great small-batch jam to make any time you find nice ripe mangoes. Plums give an interesting flavour twist to the exotic sweet–tart flavour of mangoes.

1 cup	finely chopped plums (about 3 to 4 plums)	250 mL
½ cup	water	125 mL
1	mango, peeled and diced	1
½	box (57 g) fruit pectin crystals★	½
1 tbsp	lemon juice	15 mL
3 cups	granulated sugar	750 mL

1. Place plums and water in a large stainless steel or enamel saucepan. Bring to a boil over high heat, cover, reduce heat and simmer for 10 minutes, stirring frequently.
2. Stir mango, pectin and lemon juice into plums; mix well. Return to a boil over high heat, uncovered, stirring constantly. Add sugar, return to a full boil and boil hard for 1 minute, stirring constantly. Remove from heat.
3. Ladle into sterilized jars and process as directed on page 16 (Shorter Time Processing Procedure).

Makes 4 cups (1 L).

★A half box of fruit pectin crystals measures 2 tbsp + 1 tsp or 30 mL. The remaining pectin can be used to make Microwave Winter Pear Amaretto Jam (page 45).

Lightly Brandied Nectarine Apricot Jam

Nectarines and apricots were made for each other. Put them together with a hint of brandy and they make a quite marvellous jam. Nectarines are available during the colder months, but apricots have a very short summer season. So we tried this jam using canned apricots and were very pleased with the result.

5	medium nectarines, finely chopped	5
2	cans (19 oz/540 mL) apricot halves, drained and diced	2
1 cup	water	250 mL
2 tbsp	lemon juice	25 mL
4 cups	granulated sugar	1 L
2 tbsp	brandy	25 mL

1. Measure prepared fruit (you should have about 5 cups/1.25 L).
2. Combine nectarines, apricots, water and lemon juice in a medium stainless steel or enamel saucepan. Bring to a boil over high heat, cover, reduce heat and cook for 10 minutes or until fruit is tender.
3. Add sugar to fruit, return to a boil and boil rapidly, uncovered, until mixture will form a gel,★ about 20 minutes, stirring frequently. Remove from heat and stir in brandy.
4. Ladle into sterilized jars and process as directed on page 16 (Shorter Time Processing Procedure).

Makes 6 cups (1.5 L).

★To determine when mixture will form a gel, see page 14.

Microwave Strawberry Lime Jam

You'll want to make a jar of this easy small-batch recipe when imported berries are available and you long for the taste of fresh strawberry jam. The same goes for peaches . . . see the variation below.

1½ cups	granulated sugar	375 mL
2 cups	sliced, hulled strawberries	500 mL
2 tbsp	lime juice	25 mL

1. Place sugar and strawberries in two alternating layers in a deep 12-cup (3 L) microwavable bowl. Pour lime juice over top. Do not stir.
2. Microwave, uncovered, on High (100%) for 5 minutes, stirring twice. Microwave, uncovered, on High for 10 minutes or until mixture will form a gel,★ stirring every 4 minutes.
3. Ladle into sterilized jars and process as directed on page 16 (Shorter Time Processing Procedure).

Makes 1½ cups (375 mL).

Variations:

Strawberry Lemon Jam: Replace lime juice with lemon juice.
Spiced Strawberry Jam: Tie 1 cinnamon stick, 2 whole cloves and 2 allspice berries loosely in cheesecloth and add to fruit during last 10 minutes of cooking. Remove and discard spice bag before bottling.
Peach Lime Jam: Prepare with same amount of peaches, sugar and 1 tbsp (15 mL) lime juice.

Extras to stir into any of the above variations before bottling: 1 tbsp (15 mL) finely chopped crystallized ginger; OR 1 tbsp (15 mL) Amaretto; OR ½ tsp (2 L) ground nutmeg or cinnamon.

★To determine when mixture will form a gel, see page 14.

Microwave Lemon Kiwifruit Jam

The delicate yet distinctive flavour of kiwifruit shines through refreshing lemon in this convenient-to-make microwave-cooked jam.

8	medium kiwifruit, peeled and chopped	8
1½ cups	granulated sugar	375 mL
1 tsp	grated lemon rind	5 mL
3 tbsp	lemon juice	45 mL

1. Place kiwifruit and sugar in a deep 4-cup (1 L) microwavable container. Stir in lemon rind and juice.
2. Microwave, uncovered, on High (100%) for 6 minutes, stirring twice. Microwave, uncovered, on High for 12 to 15 minutes or until mixture will form a gel,★ stirring every 4 minutes.
3. Ladle into sterilized jars and process as directed on page 16 (Shorter Time Processing Procedure).

Makes 2 cups (500 mL).

★To determine when mixture will form a gel, see page 14.

Microwave Winter Pear Amaretto Jam

The pear varieties found in winter, such as Bosc and Anjou, need the moist heat of microwave cooking to transform the relatively dry pears into a delicious jam. The almond of the liqueur highlights the taste of pears.

1½ cups	diced peeled winter pears (Bosc, Anjou)	375 mL
½ cup	chopped peeled tart apple	125 mL
2 tbsp	apple juice, cider or water	25 mL
½	box (57 g) fruit pectin crystals	½
2 cups	granulated sugar	500 mL
1 tbsp	Amaretto	15 mL

1. Combine pears, apple, apple juice and pectin in a 3-quart (3 L) microwavable container. Microwave, uncovered, on High (100%) for 6 minutes or until mixture comes to a boil, stirring twice.
2. Add sugar. Microwave, uncovered, on High for 4 minutes or until mixture returns to a full boil and boil hard for 1 minute. Stir in liqueur.
3. Ladle into sterilized jars and process as directed on page 16 (Shorter Time Processing Procedure).

Makes 3 cups (750 mL).

Tip: A half box of fruit pectin crystals measures 2 tbsp + 1 tsp or 30 mL. The remaining half can be used to make Mango Plum Jam (page 41).

Chapter Two

Jelly Made Easy

J ELLIES in their shimmering translucence are so appealing in the jar. And the good news is their minimal preparation time! Jelly making follows the same basic procedure as jam making. A properly set jelly retains its shape and quiver when removed from the jar. Some are made with a variety of fruits and vegetables — different berries, pears, apples, crabapples, plums, citrus fruits and sweet and hot peppers. And some of the very tastiest are made with wine and herbs.

Traditionally, jellies are made with the juice from cooked fruit strained through a clean jelly bag. You don't have to peel, core or remove seeds and stems before cooking, which is why jelly making is so easy and convenient. We like to press the fruit through a coarse strainer before putting it in the jelly bag. This removes the bulk of the fruit pulp, allowing the juice to strain much more quickly through the bag.

A jelly bag should be made of a clean, closely woven fabric that removes enough of the fruit pulp from the juice to ensure a clear jelly. This style of bag is available in kitchen and hardware stores. However, you can easily make a jelly bag from either clean, closely woven cheesecloth or an unused disposable all-purpose cloth. Line a colander or strainer with the cloth and set it over a larger bowl. Pour in the fruit mixture and allow it to stand until the liquid has passed through the cloth.

Some jellies use prepared juice or liquids and do not require a jelly bag. Herb and wine jellies fall into this "super-simple, quick and easy" category. A few jellies have some suspended fruit or vegetable pieces left in for enticing eye-appeal and flavour.

As discussed on pages 12–13, some fruits do not contain enough pectin to form a gel in a reasonable cooking time. Adding a commercial pectin, or using our Homemade Apple Pectin (page 35) is the solution.

Many flavourings can be added to jellies. Scented geranium leaves as well as herbs like lemon thyme, mint and angelica give exciting interest to herb jellies. Examples of this treatment are our Sherried Rosemary Grape Juice Jelly (page 49) and Basic Herb Wine Jelly (page 52). Rose petals and even fruit leaves such as peach and plum impart an almond flavour. Try adding whole herbs of your choice to wine jellies.

Serving Suggestions

Jellies have so many uses. We love the pepper jellies as appetizers with cream cheese and crackers. Our Cranberry Hot Pepper Jelly (page 54) and our wine jellies go superbly with chicken, turkey and duck. Roasted or broiled meats, whether hot or cold, are greatly enhanced by Spiced Apple Jelly (page 55) and Apple Cranberry Wine Jelly (page 50). Of course, sweet jellies such as Tangerine Lemon (page 57) and Grapefruit Raspberry Honey (page 56) go wonderfully on rolls, hot biscuits and muffins. We are sure you will find your own favourite uses. And any of our jellies make welcome gifts.

List of Recipes

Wine Jellies

Wine jelly adds a wonderful flavour note to many meals. This basic recipe can be used with sherry, port, claret or Bordeaux or, for a more delicate jelly, with white wine.

Basic Wine Jelly

2 cups	wine	500 mL
¼ cup	strained lemon juice	50 mL
3½ cups	granulated sugar	875 mL
1	pouch (85 mL) liquid fruit pectin	1

1. Place wine, lemon juice and sugar in a large stainless steel or enamel saucepan. Bring to a boil over high heat and boil hard for 1 minute, stirring constantly. Remove from heat and stir in pectin.
2. Ladle into sterilized jars and process as directed on page 16 (Shorter Time Processing Procedure).

Makes 4½ cups (1.125 L).

Cranberry Port Wine Jelly

Be sure to use pure cranberry juice. Beverages labelled "Beverage," "Drink" or "Cocktail" have been diluted and have a less-intense flavour.

1 cup	port or claret	250 mL
1 cup	pure cranberry juice	250 mL
3½ cups	granulated sugar	875 mL
1	pouch (85 mL) liquid fruit pectin	1

1. Proceed as for Basic Wine Jelly, substituting the cranberry juice for half of the wine.

Makes 4½ cups (1.125 L).

Easy Orange Port Wine Jelly

Orange juice concentrate adds a lovely intense flavour complemented by the port. For a clearer jelly, use the "pulp-free" concentrates.

1 cup	port	250 mL
1 cup	orange juice concentrate	250 mL
¼ cup	lemon juice	50 mL
3½ cups	granulated sugar	875 mL
1	pouch (85 mL) liquid fruit pectin	1

1. Proceed as for Basic Wine Jelly, substituting the orange juice concentrate for half of the wine.

Makes 4½ cups (1.125 L).

Sherried Rosemary Grape Juice Jelly

This delicate jelly has an affinity with poultry. The rosemary flavour also lends itself to pork or a leg of lamb.

1 cup	dry sherry	250 mL
1 cup	white grape juice	250 mL
3½ cups	granulated sugar	875 mL
1	pouch (85 mL) liquid fruit pectin	1
1	stem fresh rosemary, thyme or other fresh herb	1

1. Proceed as for Basic Wine Jelly, substituting grape juice for half of the wine. Add herb to jar before processing.

Makes 4½ cups (1.125 L).

Apple Cranberry Wine Jelly

This special variation of traditional cranberry jelly is quickly prepared.

2½ cups	fresh or frozen cranberries	625 mL
4	large cooking apples, peeled, cored and chopped	4
1 cup	dry white wine	250 mL
1½ cups	granulated sugar	375 mL

1. Combine cranberries, apples and wine in a large stainless steel or enamel saucepan. Bring to a boil over high heat, cover, reduce heat and cook gently for 15 minutes or until fruit is soft. Strain through a sieve, discard pulp and return sieved liquid to saucepan.
2. Add sugar and return to a boil. Boil uncovered, until mixture forms a gel★, approximately 10 minutes, stirring frequently. Remove from heat.
3. Ladle into sterilized jars and process as directed on page 16 (Shorter Time Processing Procedure).

Makes 3 cups (750 mL).

★To determine when mixture will form a gel, see page 14.

Jellies à l'Herbe

Herbs give unique twists to jellies that make wonderful gifts. Making them is much like making tea: the herbs are steeped in boiling water or wine.

Basic Herb Jelly

Beautifully simple and sparkling jelly — serve with poultry, cheese and crackers, a cold meat salad plate or sliced meat sandwiches.

1¼ cups	water	300 mL
2 tbsp	fresh herb leaves or 2 tsp (10 mL) dried	25 mL
3½ cups	granulated sugar	825 mL
1 cup	juice★	250 mL
2 tbsp	vinegar or lemon juice	25 mL
1	pouch (85 mL) liquid fruit pectin	1

1. Combine water and herb leaves in a small saucepan. Bring mixture to a boil; remove from heat. Cover and allow to steep for 5 minutes. Strain through a lined sieve; discard leaves.
2. Place 1 cup (250 mL) of the liquid in a large stainless steel or enamel saucepan; add sugar, juice and vinegar or lemon juice. Bring to a full boil over high heat and boil hard for 1 minute, stirring constantly. Remove from heat and stir in pectin.
3. Ladle into sterilized jars and process as directed on page 16 (Shorter Time Processing Procedure).

Makes about 4½ cups (1.125 L).

★ Many juices may be used in making herb jellies. Match the herb and the vinegar to the juice and proceed as above. Some examples are:
- Grape juice, thyme and lemon juice
- Tomato juice, basil and red wine vinegar
- Orange juice, basil and rice vinegar
- White grape juice, sage and cider vinegar
- Apple juice, dried thyme, savoury, marjoram or oregano and cider vinegar

Basic Herb Wine Jelly

Use white wine if the jelly is to accompany poultry or pork, or red wine if it is to accompany red meats. Either way, it's a sparkling clear jelly enhanced with the flavour of your favourite herb. Some herbs to try are sage, basil, mint, savoury, thyme, marjoram and oregano. The amount used may be increased or decreased according to individual taste preferences.

1¾ cups	dry white or red wine	425 mL
¼ cup	white or red wine vinegar	50 mL
3 tbsp	fresh herb leaves or 1 tbsp (15 mL) dried	45 mL
3½ cups	granulated sugar	875 mL
1	pouch (85 mL) liquid fruit pectin	1

1. Combine wine, vinegar and herb leaves in a large stainless steel or enamel saucepan. Bring mixture to a boil; remove from heat. Cover and allow to steep for 30 minutes to extract flavours.
2. Strain mixture through a lined sieve; discard leaves. Return liquid to saucepan and stir in sugar. Bring to a boil over high heat and boil rapidly for 1 minute, stirring constantly. Remove from heat and stir in pectin.
3. Ladle into sterilized jars and process as directed on page 16 (Shorter Time Processing Procedure).

Makes 4 cups (1 L).

Tip: Add several garlic cloves to Basic Herb Wine Jelly made with white wine and rosemary for a perfect jelly to serve with a crown roast of lamb or lamb chops.

Sparkling Sweet Pepper Jelly

Pieces of red, yellow and orange pepper sparkle like jewels in this exotic jelly. Our favourite way to serve it is with cream cheese spread on melba toast rounds. We also like it with a sliver of Cheddar cheese on a cracker. If you like your pepper jelly hot, see the variation below.

½ cup	each evenly diced sweet red, orange and yellow pepper	125 mL
¾ cup	white wine vinegar	175 mL
3 cups	granulated sugar	750 mL
1	pouch (85 mL) liquid fruit pectin	1

1. Combine peppers, vinegar and sugar in a large stainless steel or enamel saucepan. Bring to a full boil over high heat and boil hard for 1 minute, stirring constantly. Add pectin; return to boil and boil rapidly for 1 minute. Remove from heat.

2. Ladle into sterilized jars and process as directed on page 16 (Shorter Time Processing Procedure).

Makes 3½ cups (875 mL).

Variations:

Sparkling Hot Pepper Jelly: Use 2 jalapeño peppers, seeded and diced, to replace the yellow or orange pepper.

Sparkling Apricot Hot Pepper Jelly: Use 2 jalapeño peppers, seeded and diced, to replace the yellow pepper and use ¼ cup (50 mL) chopped dried apricots, soaked in water for 4 to 6 hours, to replace the orange pepper.

Cranberry Hot Pepper Jelly

The colour is as intense as the flavour of this sparkling red jelly. Adding cranberry to popular pepper jelly makes it perfect for the festive season. Keep it on hand for an easy appetizer with cream cheese on crackers, or to accompany any roast meat, especially game such as venison. Our version has medium heat, but you may easily change it by adding or omitting a jalapeño pepper.

1	large sweet red pepper	1
2	jalapeño peppers, seeded, or other hot pepper	2
¼ cup	water	50 mL
¾ cup	cider vinegar	175 mL
¾ cup	frozen cranberry cocktail concentrate, thawed	175 mL
3 cups	granulated sugar	750 mL
1	pouch (85 mL) liquid fruit pectin	1

1. Finely chop sweet and jalapeño peppers in food processor. Place in a small saucepan with water and vinegar. Bring mixture to a boil, cover, reduce heat and boil gently for 10 minutes. Strain mixture through a coarse sieve, pressing with back of a spoon to extract as much liquid as possible; discard solids. Pour liquid through a jelly bag.
2. Place strained liquid, cranberry concentrate and sugar in a large stainless steel or enamel saucepan. Bring to a full boil over high heat, stirring constantly. Stir in pectin, return to a full boil and boil hard for 1 minute, stirring constantly. Remove from heat.
3. Ladle into sterilized jars and process as directed on page 16 (Shorter Time Processing Procedure).

Makes 3 cups (750 mL).

Variation:

Cranberry Hot Pepper Jelly with Balsamic Vinegar: Replace ¼ cup (50 mL) cider vinegar with balsamic vinegar.

Spiced Apple Jelly

We adapted this recipe from one Margaret's neighbour makes every fall using her own apples. Don't peel, core or remove the seeds and stems before cooking; they add pectin as well as colour and flavour to the jelly. The crabapple variation is wonderful with poultry.

2 lb	apples, cut into large pieces (about 8 cups/2 L)	2 L
6 cups	water	1.5 L
¼ cup	cider vinegar	50 mL
½ tsp	whole cloves	2 mL
2	cinnamon sticks, 3 inches (7.5 cm) long	2
1½ cups	granulated sugar	375 mL

1. Combine apples, water, vinegar, cloves and cinnamon sticks in a large stainless steel or enamel saucepan. Cover and bring to a boil over high heat, reduce heat and boil gently for 30 minutes. Strain mixture through a coarse sieve; discard solids. Pour liquid through a jelly bag.
2. Return strained liquid to pan and add sugar. Bring to a boil and boil rapidly, uncovered, until mixture will form a gel,★ about 15 minutes, stirring occasionally. Remove from heat.
3. Ladle into sterilized jars and process as directed on page 16 (Shorter Time Processing Procedure).

Makes 2 cups (500 mL).

Variation:

Crabapple Jelly: Replace apples with the same weight of crabapples.

★To determine when mixture will form a gel, see page 14.

Grapefruit Raspberry Honey Jelly

With the bright colour of raspberry and hint of honey, this jelly will become an instant favourite.

2	grapefruit, coarsely chopped	2
3 cups	water	750 mL
3 cups	fresh or frozen unsweetened raspberries	750 mL
3 cups	granulated sugar	750 mL
1 cup	liquid honey	250 mL

1. Place grapefruit and water in a medium stainless steel or enamel saucepan. Bring to a boil over high heat, cover, reduce heat and boil gently for 20 minutes. Add raspberries, return to a boil and boil gently for 5 minutes.
2. Pour fruit through a coarse strainer, pressing pulp to extract juice. Discard solids. Pour liquid through a jelly bag.
3. Return strained liquid to pan and add sugar and honey. Bring to a boil and boil rapidly, uncovered, until mixture will form a gel,★ about 15 minutes, stirring occasionally. Remove from heat.
4. Ladle into sterilized jars and process as directed on page 16 (Shorter Time Processing Procedure).

Makes 2 cups (500 mL).

★To determine when mixture will form a gel, see page 14.

Tangerine Lemon Jelly

We love this jewel-like jelly on hot biscuits or with cream cheese and bagels. For a special gift, add a sprig of fresh rosemary to the jelly before processing.

3	lemons	3
9–10	tangerines	9–10
1	box (57 g) fruit pectin crystals	1
4½ cups	granulated sugar	1.125 L

1. Squeeze lemons and tangerines to give 4 cups (1 L) juice. Bring juice to a boil over high heat in a large stainless steel or enamel saucepan, cover, reduce heat and simmer for 10 minutes, stirring occasionally. Remove from heat.
2. Strain juice through a jelly bag. Return strained liquid to saucepan and stir in pectin. Bring to a full boil over high heat, reduce heat and boil gently for 1 minute, stirring constantly. Add sugar, return to a full boil and boil hard for 1 minute, stirring constantly. Remove from heat.
3. Ladle into sterilized jars and process as directed on page 16 (Shorter Time Processing Procedure).

Makes 5½ cups (1.375 L).

Tip: To obtain a very clear jelly without a jelly bag, strain the liquid through a lined sieve. We have found disposable cloth works really well for this.

Chapter Three

Marvellous Marmalades

MARMALADES are jams made from citrus fruit and their peel — oranges, lemons, grapefruit, limes and tangerines. The word *marmalade* originates from *marmalada*, a Portuguese word for preserved quinces, enjoyed in medieval times. Today's marmalades frequently contain other fruits in combination with the citrus fruits. Since citrus fruits are so rich in pectin (especially in the rind and seeds), they rarely require any added pectin to form a gel. But marmalades may require a longer time for the gel to form.

The very best time to make most marmalades is during the winter when citrus fruits are at their best quality and best price. The bitter oranges, Seville-type being one, are available only for a short time, around late January and into February. However, we have discovered how to prolong this short season: freeze the oranges whole and then make our superb Traditional English Seville Marmalade (page 60) when time permits.

Marmalades do take a bit of time, but are well worth the effort. Citrus zesters, vegetable peelers, juice extractors and food processors take much of the labour out of their preparation. We are really quite excited about the marmalades in this chapter. Not all marmalades are bitter-sweet, as many people think. They range from a sweet Cherry Brandy (page 69) to an Old-Fashioned Tomato (page 66) that even Grandma would be proud of. And for marmalades a bit more upbeat, we like Mango (page 68) and Lemon Ginger Zucchini (page 67).

Serving Suggestions

Marmalades are not just for spreading on breakfast toast, although this is still by far their greatest use. We offer several recipes in chapter 14 using any citrus marmalade — look for an easy fruitcake made with any marmalade (page 60-69).

List of Recipes

Traditional English Seville Marmalade

Don't be put off by the taste of fresh Seville-type oranges. Because of its high acid content, the Seville is not an eating orange. Yet its bitterness is magically transformed into a traditional English-style marmalade. Seville-type oranges are generally available in January and February, so mark your calendar to make a batch or two to enjoy throughout the year.

4	Seville-type oranges	4
2	lemons, very thinly sliced	2
4 cups	water	1 L
¼ tsp	baking soda	1 mL
4 cups	granulated sugar	1 L

1. Remove thin outer rind from oranges with vegetable peeler and cut into fine strips with scissors or sharp knife; or use a zester. Place rind, lemons and water in a large stainless steel or enamel saucepan. Bring to a boil over high heat, cover, reduce heat and boil gently for 25 minutes, stirring occasionally.
2. Remove and discard remaining white rind from oranges. Cut oranges in half. Working over a bowl to catch juices, remove seeds with a sharp knife or fork. Place orange halves and juice in a food processor or blender and process until finely chopped.
3. Add chopped pulp and baking soda to lemon mixture in saucepan. Bring to a boil over high heat, reduce heat, cover and boil gently for 20 minutes, stirring frequently.
4. Add sugar to fruit mixture. Return to a boil and boil rapidly, uncovered, until mixture will form a gel,★ about 20 minutes, stirring frequently. Remove from heat.
5. Ladle into sterilized jars and process as directed on page 16 (Shorter Time Processing Procedure).

Makes about 6½ cups (1.625 L).

   Tip: If you are using frozen Seville-type oranges, remove the rind from the orange while it is still frozen.

   ★To determine when mixture will form a gel, see page 14.

Ruby-Red Grapefruit Marmalade

Eat your grapefruit on your toast? You certainly can with this grapefruit marmalade. The ruby-red fruit makes a delicious and attractive delicate pink marmalade.

3	pink grapefruit	3
2	lemons	2
3 cups	water	750 mL
3½ cups	granulated sugar	875 mL

1. Remove thin outer rind from grapefruit and lemons with vegetable peeler and cut into fine strips with scissors or sharp knife; or use a zester. Place rind and water in a large stainless steel or enamel saucepan. Bring to a boil over high heat; cover, reduce heat and boil gently for 20 minutes.
2. Remove and discard remaining white rind and seeds from fruit. Finely chop pulp in a food processor or blender and add to saucepan. Bring to a boil over high heat, reduce heat, cover and boil gently for 10 minutes, stirring frequently.
3. Add sugar to fruit. Return to a boil over high heat and boil rapidly, uncovered, until mixture will form a gel,★ about 30 minutes, stirring frequently. Remove from heat.
4. Ladle into sterilized jars and process as directed on page 16 (Shorter Time Processing Procedure).

Makes 4½ cups (1.125 L).

Suggestion:

Spirited Marmalades: Liqueurs, rum, brandy, whiskey or nuts turn marmalade into an elegant and luscious breakfast preserve. When the mixture will form a gel, add 2 tbsp (25 mL) of the chosen spirit or nuts and cook 5 minutes longer before bottling.

★To determine when mixture will form a gel, see page 14.

Cranberry Orange Marmalade

The shiny scarlet cranberry lends a tartness to complement the orange of this marmalade. Make this marmalade when cranberries are at their peak between Thanksgiving and Christmas. But since cranberries freeze well, you can also make this marmalade year round.

2	medium oranges	2
1	lemon	1
3 cups	water	750 mL
2 cups	fresh or frozen cranberries	500 mL
4 cups	granulated sugar	1 L

1. Remove thin outer rind from oranges and lemon with vegetable peeler and cut into very fine strips with scissors or sharp knife; or use a zester. Place rind and water in a large stainless steel or enamel saucepan. Bring to a boil over high heat, cover, reduce heat and boil gently for 20 minutes.
2. Remove and discard remaining white rind and seeds from oranges and lemon. Finely chop pulp and cranberries in a food processor or blender and add to saucepan. Bring to a boil over high heat; reduce heat, cover and boil gently for 10 minutes, stirring occasionally.
3. Add sugar to fruit mixture. Return to a boil over high heat and boil rapidly, uncovered, until mixture will form a gel,★ about 20 minutes, stirring frequently. Remove from heat.
4. Ladle into sterilized jars and process as directed on page 16 (Shorter Time Processing Procedure).

Makes about 5 cups (1.25 L).

★To determine when mixture will form a gel, see page 14.

Fresh Pineapple Marmalade with Lemon

Be sure your pineapple is fully ripe to best enjoy the flavour of this wonderful marmalade. It should be slightly soft to the touch with a strong colour and no sign of green.

2	lemons	2
2 cups	chopped fresh pineapple, peeled and cored (about ½ pineapple)	500 mL
2½ cups	water	625 mL
3 cups	granulated sugar	750 mL

1. Remove thin outer rind from lemons with vegetable peeler and cut into fine strips with scissors or sharp knife; or use a zester. Place in a large stainless steel or enamel saucepan. Remove and discard remaining white rind and seeds.
2. Finely chop lemon and pineapple in a food processor or blender. Add fruit and water to saucepan. Bring to a boil over high heat, cover, reduce heat and boil gently for 20 minutes, stirring frequently.
3. Add sugar to saucepan, return to a boil over high heat and boil rapidly, uncovered, until mixture will form a gel,★ about 35 minutes, stirring frequently. Remove from heat.
4. Ladle into sterilized jars and process as directed on page 16 (Shorter Time Processing Procedure).

Makes 3½ cups (875 mL).

★To determine when mixture will form a gel, see page 14.

Pear Kiwifruit Lime Marmalade

The brilliant green flesh of the kiwifruit gives this intriguing marmalade its beautiful jewel-like colour. The principal growers of kiwifruit, New Zealand and California, have opposite growing seasons, giving us virtually year-round availability.

3	limes, very thinly sliced	3
2½ cups	water	625 mL
2 cups	chopped peeled and cored pears (about 3 pears)	500 mL
2 cups	chopped peeled kiwifruit (about 9 kiwifruit)	500 mL
4½ cups	granulated sugar	1.125 L

1. Place limes and water in a large stainless steel or enamel saucepan. Bring to a boil over high heat, cover, reduce heat and boil gently for 20 minutes, stirring occasionally.
2. Add pears and kiwifruit. Bring to a boil over high heat, cover, reduce heat and boil gently for 10 minutes, stirring occasionally.
3. Add sugar, return to a boil over high heat and boil rapidly, uncovered, until mixture will form a gel,★ about 35 minutes, stirring frequently. Remove from heat.
4. Ladle into sterilized jars and process as directed on page 16 (Shorter Time Processing Procedure).

Makes about 6½ cups (1.625 L).

★To determine when mixture will form a gel, see page 14.

Five-Fruit Marmalade

In early winter when honey tangerines, grapefruit, lemons, limes and sweet oranges are at their best, make this tangy variation of a traditional marmalade. Don't just eat it at breakfast. Try it as a glaze on chicken breasts, baked ham and roasted pork.

2	lemons	2
2	limes	2
2–3	medium oranges	2–3
1	grapefruit	1
2	tangerines, peeled	2
4 cups	water	1 L
¼ tsp	baking soda	1 mL
5½ cups	sugar	1.375 L

1. Remove thin outer rind from lemons, limes, 2 oranges and grapefruit with vegetable peeler and cut into fine strips with scissors or sharp knife; or use a zester. Place in a large stainless steel or enamel saucepan. Remove the white rind in large pieces from lemons, oranges and grapefruit and place in saucepan. Add water; bring to a boil over high heat, cover, reduce heat and boil gently for 25 minutes.
2. Remove and discard remaining white rind from limes. Finely chop all fruit pulp in a food processor or blender; it should measure 4 cups (1 L). (Add the chopped pulp of the remaining orange if needed.) Add fruit and baking soda to saucepan. Bring to a boil over high heat, cover, reduce heat and boil gently for 20 minutes, stirring frequently. Using tongs, remove and discard the large pieces of rind.
3. Add sugar to saucepan and return to a boil, stirring constantly. Boil rapidly, uncovered, until mixture will form a gel,★ about 30 minutes, stirring frequently. Remove from heat.
4. Ladle into sterilized jars and process as directed on page 16 (Shorter Time Processing Procedure).

Makes about 6 cups (1.5 L).

Tip: Use Five-Fruit Marmalade to make Marmalade Squares (page 225).

★   To determine when mixture will form a gel, see page 14.

Old-Fashioned Tomato Marmalade

Tomatoes impart a delicate fresh flavour to this preserve that is unique among marmalades. It's John Howard's family recipe. A variation adds chopped ginger-root during the cooking for a marmalade that is extra-special served with chicken, pork or fish.

5 cups	coarsely chopped peeled tomatoes (about 2½ lb/2 kg)	1.25 L
2	large oranges	2
1	lemon	1
4 cups	granulated sugar	1 L

1. Place tomatoes in a large stainless steel or enamel saucepan.
2. Halve and seed oranges and lemon. Finely chop fruit in food processor or blender and add to tomatoes. Bring mixture to a full boil over high heat. Slowly add sugar, stirring until sugar is completely dissolved. Return to a boil and boil rapidly until mixture will form a gel,★ about 1 hour, stirring frequently. Remove from heat.
3. Ladle into sterilized jars and process as directed on page 16 (Shorter Time Processing Procedure).

Makes about 6 cups (1.5 L).

Variation:

Gingered Tomato Marmalade: Add 3 tbsp (45 mL) finely chopped peeled gingerroot during cooking.

★To determine when mixture will form a gel, see page 14.

Lemon Ginger Zucchini Marmalade

Fresh ginger combines with lemon to give a magnificent zing to this marmalade. It's a nice change from the sweeter types. Chop the ginger finely for a stronger flavour.

3	lemons	3
1	medium orange	1
2½ cups	water	625 mL
½ cup	chopped fresh peeled gingerroot	125 mL
1 cup	shredded zucchini	250 mL
4½ cups	granulated sugar	1.125 L

1. Remove thin outer rind from lemons and orange with vegetable peeler and cut into fine strips with scissors or sharp knife; or use a zester. Place in a large stainless steel or enamel saucepan. Remove the remaining white rind in large pieces and add to saucepan. Stir in water and gingerroot. Bring to a boil over high heat, cover, reduce heat and boil gently for 25 minutes. Using tongs, remove and discard white rind.
2. Finely chop fruit pulp in a food processor or blender. Add pulp and zucchini to saucepan. Bring to a boil over high heat, reduce heat, cover and boil gently for 20 minutes, stirring occasionally.
3. Add sugar to fruit mixture. Return to a boil and boil rapidly, uncovered, until mixture will form a gel,★ about 30 minutes, stirring frequently.
4. Ladle into sterilized jars and process as directed on page 16 (Shorter Time Processing Procedure).

Makes about 4½ cups (1.125 L)

★To determine when mixture will form a gel, see page 14.

Mango Marmalade

The exotic sweet–tart flavour of mango permeates this tropical marmalade.

2	lemons, very thinly sliced	2
2 cups	water	500 mL
2	mangoes, peeled and thinly sliced	2
2 cups	granulated sugar	500 mL

1. Combine lemons and water in a large stainless steel or enamel saucepan. Bring to a boil over high heat, cover, reduce heat and boil gently for 25 minutes, stirring occasionally.
2. Add mangoes to saucepan. Bring to a boil over high heat, stirring constantly; reduce heat, cover and boil gently for 20 minutes, stirring occasionally.
3. Stir in sugar. Return to a boil and boil rapidly, uncovered, until mixture will form a gel,★ about 15 minutes, stirring frequently.
4. Ladle into sterilized jars and process as directed on page 16 (Shorter Time Processing Procedure).

Makes about 3 cups (750 mL).

★To determine when mixture will form a gel, see page 14.

Cherry Brandy Marmalade

Sour cherries have a definite affinity with citrus fruits, offering a wonderful fresh flavour.

2	medium oranges, halved and thinly sliced	2
1 cup	water	250 mL
3 cups	fresh or frozen sour cherries	750 mL
2 tbsp	lemon juice	25 mL
3½ cups	granulated sugar	875 mL
2 tbsp	brandy (optional)	25 mL

1. Place orange slices and water to barely cover in a medium stainless steel or enamel saucepan. Bring to a boil over high heat, cover, reduce heat and cook for 10 minutes or until fruit is tender.
2. Add cherries and lemon juice to saucepan. Return to a full boil and slowly add sugar. Boil rapidly, uncovered, until mixture will form a gel,★ about 20 minutes, stirring occasionally. Remove from heat and stir in brandy (if using).
3. Ladle into sterilized jars and process as directed on page 16 (Shorter Time Processing Procedure).

Makes 3½ cups (875 mL).

Variation:

Fresh herbs may be added to many marmalades. For example, replacing the brandy with 1 tbsp (15 mL) fresh mint leaves will lend a new flavour to this cherry marmalade.

★To determine when mixture will form a gel, see page 14.

Chapter Four

Conserves, Butters and Curds

ONSERVES are best described as cousins to jams. Usually nuts and dried fruits are added and frequently more than one fruit is used. Most of the summer fruits, cherries and blueberries, as well as the fall cranberries, apples, pears and plums lend themselves well to conserves. In an earlier era conserves were often eaten as a dessert. Today, they are more commonly enjoyed as either a dessert sauce or a spread. And in some cases they are served as a savoury accompaniment to meats; Apricot Grand Marnier Conserve (page 72) is a delicious example.

Generally, conserves are made the same way as jams. Since they are lower in pectin, they form a lighter gel. When dried fruits are called for, it is best to soak them for a few hours or overnight; they will swell, become softer and give a much better final yield after soaking.

Whereas we have advised using slightly underripe fruit in jam and jelly, softer, fully sun-ripened fruits are the best choice for fruit butters. These fruits are often available at lower prices. Fruit butters are very easy to prepare. They are smooth, yummy and spreadable.

Fruit curds, though less well known than other forms of sweet spreads, are no less tasty or useful in any creative kitchen. They are a butter-and-eggs enrichment of zesty fresh citrus fruits. Microwave Lemon Curd (page 82) is by far the most common, and we offer several variations. Our microwave method makes fool-proof success just a short "zap" away. It is best to make fruit curds in small amounts to store in half-pint (250 mL) jars. Curds may be kept refrigerated for 3 weeks, but since their fresh taste fades, they are better eaten sooner than later.

Freeze them for extended storage, keeping them ready for a fast defrost before using as an easy dessert.

Serving Suggestions

We think conserves are among the very best items to "put a lid on" for gift giving. They offer a special touch of "luxury" that makes them a bit more special than other spreads. Fruit curds make sublimely delicious desserts. Use them as a cake filling or icing, as a tart or pie filling, with angel cake, meringues and lady fingers or as a base for fresh fruits. Like conserves, fruit curds make great gifts. Fruit butters make a marvellous substitute for high-fat spreads. They are also excellent for low-fat baking — see Sweet and Chunky Apple Butter Spice Cake (page 228) and Spiced Plum Butter Bran Muffins (page 213). Served with nippy cheeses, like Cheddar or Stilton, and plain crackers, fruit butters become an easy snack or an elegant dessert. Some make fine accompaniments to savoury dishes.

List of Recipes

Apricot Grand Marnier Conserve

A beautiful golden colour, this conserve follows through with great apricot and orange flavours. Use as a savoury accompaniment to meats, or to top piping-hot biscuits or crispy herb toast. For a simple yet splendid main dish, try it in Chicken with Apricot Sauce (page 224).

2 cups	diced dried apricots	500 mL
4½ cups	water	1.125 L
1	large tart apple, peeled, cored and chopped	1
1 tsp	finely grated lemon rind	5 mL
¼ cup	lemon juice	50 mL
4 cups	granulated sugar	1 L
⅓ cup	Grand Marnier OR orange juice concentrate	75 mL
½ cup	slivered almonds	125 mL

1. Place apricots and water in a large stainless steel or enamel saucepan. Cover and let stand for at least 4 hours or overnight.
2. Add apple, lemon rind and lemon juice. Bring to a full boil over high heat, reduce heat, cover and boil gently for about 15 minutes or until fruit is tender, stirring occasionally.
3. Add sugar to saucepan. Return to a boil, reduce heat and boil gently, uncovered, until mixture will form a light gel,★ about 25 minutes, stirring occasionally.
4. Add liqueur, return to a boil and boil gently for 5 minutes. Remove pan from heat and stir in almonds.
5. Ladle into sterilized jars and process as directed on page 16 (Shorter Time Processing Procedure).

Makes 5 cups (1.25 L).

Tip: The dried apricots are soaked in water to soften them and obtain a better cooked yield.

★To determine when mixture will form a light gel, see page 14.

Winter Dried Fruit and Nut Conserve

You will find this marvellous conserve fabulous with goose, turkey or chicken, duck and pork roasts. It reminds us of mincemeat, but for meats. We've been known to eat it with cheese and crackers also.

2	Granny Smith apples, peeled, cored and diced	2
2	winter pears, peeled, cored and diced	2
½ cup	finely chopped dates	125 mL
½ cup	raisins	125 mL
½ cup	dried cranberries	125 mL
½ cup	apple juice	125 mL
2 cups	lightly packed brown sugar	500 mL
3 tbsp	lemon juice	45 mL
½ cup	coarsely chopped pecans	125 mL
Pinch	each ground allspice, nutmeg and ginger	Pinch

1. Place apples, pears, dates, raisins, cranberries and apple juice in a large stainless steel or enamel saucepan. Bring to a full boil over high heat, cover, reduce heat and boil gently for 10 minutes or until fruit is tender, stirring occasionally.
2. Stir in sugar and lemon juice. Return to a boil, reduce heat and boil gently, uncovered, until mixture will form a light gel,★ about 10 minutes, stirring frequently. Remove from heat and stir in nuts, allspice, nutmeg and ginger.
3. Ladle into sterilized jars and process as directed on page 16 (Shorter Time Processing Procedure).

Makes 3½ cups (875 mL).

★To determine when mixture will form a light gel, see page 14.

Gingered Pear Apricot Conserve

An adventure in fruit, nut and spice flavours, this conserve is great as a tart filling or over ice cream.

1	large lime	1
4 cups	finely chopped peeled and cored pears (4 large pears)	1 L
½ cup	water	125 mL
2½ cups	granulated sugar	625 mL
½ cup	chopped dried apricots	125 mL
¼ cup	finely chopped crystallized ginger	50 mL
¼ cup	slivered almonds	50 mL

1. Remove thin outer rind from lime with vegetable peeler and cut into fine strips with scissors or sharp knife; or use a zester. Remove and discard remaining white rind. Finely chop lime pulp with a knife or in a food processor with on/off motion. Place lime rind and pulp in a large stainless steel or enamel saucepan; add pears and water. Bring to a boil over high heat, cover and boil gently for 10 minutes or until fruit is tender.
2. Stir in sugar, apricots and ginger. Return to a boil, reduce heat and boil gently, uncovered, until mixture will form a light gel,★ about 20 minutes, stirring occasionally. Remove from heat and stir in almonds.
3. Ladle into sterilized jars and process as directed on page 16 (Shorter Time Processing Procedure).

Makes 4 cups (1 L).

★To determine when mixture will form a light gel, see page 14.

Maple Blueberry Conserve with Walnuts

Any kind of blueberries, especially wild ones, make a marvellous conserve. Combine this with the nectar of the maple tree and you have a real Canadian treat. Fold it into yogurt for a pancake or waffle topping, or just spread it on toast or muffins.

2 cups	fresh or frozen blueberries, crushed	500 mL
½ cup	water	125 mL
¼ cup	maple syrup	50 mL
1 tbsp	lemon juice	15 mL
1 cup	granulated sugar	250 mL
½ cup	raisins	125 mL
¼ cup	chopped walnuts	50 mL
½ tsp	each ground allspice and ginger	2 mL

1. Combine blueberries, water, maple syrup and lemon juice in a large stainless steel or enamel saucepan. Bring to a boil over high heat, cover, reduce heat and boil gently for about 5 minutes or until fruit is tender, stirring occasionally.
2. Stir in sugar and raisins. Return to a boil, reduce heat and boil gently, uncovered, until mixture will form a light gel,★ about 15 minutes, stirring occasionally. Remove from heat and stir in walnuts, allspice and ginger.
3. Ladle into sterilized jars and process as directed on page 16 (Shorter Time Processing Procedure).

Makes 1½ cups (375 mL).

★ To determine when mixture will form a light gel, see page 14.

Blueberry Honey
Another blueberry pleasure for morning toast.
Finely chop ½ cup (125 mL) blueberries in a food processor. Add 1 cup (250 mL) creamed honey and pulse until blended. Store in a tightly sealed container. Makes 1⅓ cups (325 mL).

Sour Cherry Hazelnut Conserve

Cherry pieces and hazelnut halves suspended in a ruby-red gel promise the rich flavours to come. Enjoy this conserve with toasted crumpets or English muffins at teatime.

4 cups	coarsely chopped pitted fresh or frozen sour cherries (about 6 cups/1.5 L whole)	1 L
1	lemon	1
1	medium orange	1
⅔ cup	dry white wine	150 mL
3 cups	granulated sugar	750 mL
½ cup	halved hazelnuts (filberts)	125 mL

1. Place chopped cherries in a large stainless steel or enamel saucepan.
2. Remove thin outer rind from lemon and orange with vegetable peeler or zester, chop finely and add to saucepan. Remove and discard remaining white rind from lemon and orange; chop pulp into small pieces. Add pulp and wine to cherries. Bring to a boil over high heat, cover, reduce heat and simmer for 10 minutes or until fruit is tender.
3. Stir in sugar. Return to a boil, reduce heat and boil gently, uncovered, until mixture will form a light gel,★ about 20 minutes, stirring occasionally. Remove pan from heat and stir in nuts.
4. Ladle into sterilized jars and process as directed on page 16 (Shorter Time Processing Procedure).

Makes 4 cups (1 L).

Lemon-Scented Honey
A refreshing and tangy fruit honey with tea biscuits and toasted English muffins.
Combine ½ cup (125 mL) creamed honey, 2 tbsp (25 mL) lemon juice and 1 tsp (5 mL) grated lemon rind in a small saucepan. Heat briefly, stirring well. Store in a tightly sealed container. Makes ½ cup (125 mL).

★To determine when mixture will form a light gel, see page 14.

Kiwifruit Honey Almond Conserve

Sliced almonds suspended in an emerald gel entice us to taste this luxurious teaming of kiwifruit, honey and almonds. Amaretto adds a nice flavour note and a softer texture.

1½ cups	diced peeled kiwifruit	375 mL
1 cup	diced peeled and cored apple	250 mL
1	lemon	1
¼ cup	water	50 mL
⅔ cup	granulated sugar	150 mL
⅔ cup	liquid honey	150 mL
½ cup	raisins	125 mL
½ cup	sliced almonds	125 mL
2 tbsp	Amaretto (optional)	25 mL

1. Place kiwifruit and apple in a large stainless steel or enamel saucepan.
2. Remove thin outer rind from lemon with vegetable peeler and cut into fine strips with scissors or sharp knife; or use a zester and add to saucepan. Remove and discard the remaining white rind. Finely chop the pulp with a knife or in a food processor with on/off motion. Add pulp and water to saucepan. Bring to a boil over high heat, cover, reduce heat and boil gently for 10 minutes or until fruit is tender.
3. Stir in sugar, honey and raisins. Return to a boil, reduce heat and boil gently, uncovered, until mixture will form a light gel,★ about 25 minutes, stirring occasionally. Remove from heat; stir in nuts and liqueur (if using).
4. Ladle into sterilized jars and process as directed on page 16 (Shorter Time Processing Procedure).

Makes 3 cups (750 mL).

★To determine when mixture will form a light gel, see page 14.

Cranberry Port Conserve

A good friend, Jane Hope, a Toronto home economist, inspired this recipe using one of our favourite festive-season fruits, cranberries. It makes a dandy gift any time of the year for serving with hot tea biscuits, game or poultry. Jane passed away a few years ago. We remember her again with this wonderful recipe.

4 cups	fresh or frozen cranberries	1 L
2 cups	granulated sugar	500 mL
¾ cup	port	175 mL
½ cup	finely chopped peeled orange	125 mL
⅓ cup	raisins	75 mL
¼ cup	chopped walnuts	50 mL

1. Combine cranberries, sugar and port in a large stainless steel or enamel saucepan. Bring to a full boil over high heat and cook, uncovered, until berries pop.
2. Add orange and raisins. Return to a boil, reduce heat and boil gently, uncovered, until mixture will form a light gel,★ about 15 minutes, stirring occasionally. Remove from heat and stir in nuts.
3. Ladle into sterilized jars and process as directed on page 16 (Shorter Time Processing Procedure).

Makes 4 cups (1 L).

Raspberry Honey with Chambord
A delightful fruit honey with toast.
Stir together well ¼ cup (50 mL) sieved fresh or frozen unsweetened raspberries, 1 tbsp (15 mL) Chambord or Raspberry Schnapps and 1 cup (250 mL) creamed honey. Store in a tightly sealed container. Makes 1 cup (250 mL).

Tip: Cranberry Port Conserve makes a delicious appetizer cheese spread (page 216).

★To determine when mixture will form a light gel, see page 14.

Sweet and Chunky Apple Butter

This fruit butter makes a quick dessert. It's also a great snack on bread or toast. We use it in a low-fat recipe — a moist Sweet and Chunky Apple Butter Spice Cake (page 228). We have found preserving in half-pint (250 mL) jars convenient, since these recipes call for that amount of apple butter. But if you use larger jars, you'll have lots left for other uses.

2 lb	McIntosh apples, peeled and cored (6 large apples)	1 kg
2 lb	Granny Smith apples, peeled and cored (4 large apples)	1 kg
1 cup	apple cider	250 mL
2 cups	granulated sugar	500 mL
2 tbsp	lemon juice	25 mL

1. Cut McIntosh apples into 1-inch (2.5 cm) pieces. Cut Granny Smith apples into smaller dice.
2. Combine apples and cider in a large stainless steel or enamel saucepan. Bring to a boil over medium-high heat, stirring occasionally. Reduce heat and boil gently for 20 minutes or until mixture is reduced by half.
3. Stir in sugar and lemon juice. Return to a boil, reduce heat and boil gently for about 25 minutes or until mixture is very thick. There should still be some tender apple chunks remaining. Remove from heat.
4. Ladle into sterilized jars and process as directed on page 16 (Shorter Time Processing Procedure).

Makes 7 cups (1.75 L).

Variation:

Spiced Apple Butter: Add 2 tsp (10 mL) ground cinnamon and ½ tsp (2 mL) each ground cloves and allspice with the sugar.

Tip: Sweet and Chunky Apple Butter adds lots of flavour to Nippy Apple Cheddar Soup (page 219).

Spiced Plum Butter

The *Shorter Oxford Dictionary* defines plum as a "good thing . . . the pick or best of a collection of things." This is an apt reflection of the high esteem held for the plum fruit. We think this recipe is a plum among butters. It works equally well with blue, red or purple plums.

10	plums, sliced	10
1 cup	water	250 mL
	granulated sugar	
1	cinnamon stick	1
4	whole cloves	4
½ tsp	ground nutmeg (optional)	2 mL

1. Place plums and water in a large stainless steel or enamel saucepan. Bring to a full boil over high heat, cover, reduce heat and simmer for 20 minutes or until plums are tender, stirring occasionally.
2. Place plum mixture in a food processor or blender and process until almost smooth. Measure and return to saucepan. For each 1 cup (250 mL) plums, add 1¼ cups (300 mL) sugar. Tie cinnamon and cloves in a spice bag and add to saucepan.
3. Return plum mixture to a boil, reduce heat and boil gently, uncovered, until mixture is very thick, stirring frequently.
4. Discard spice bag; stir in nutmeg (if using).
5. Ladle into sterilized jars and process as directed on page 16 (Shorter Time Processing Procedure).

Makes about 3 cups (750 mL).

Variation:

Spiced Pear Butter: Replace plums with 5 peeled, cored and sliced pears.

Tip: Use this butter to make Spiced Plum Butter Bran Muffins (page 213).

Apricot Honey Butter

Spread this elegant ambrosia on English muffins, pancakes or waffles. You'll never miss butter again!

2 cups	chopped dried apricots	500 mL
2 tbsp	grated lemon rind	25 mL
2 cups	water	500 mL
½ cup	lemon juice	125 mL
¼ cup	finely chopped crystallized ginger	50 mL
⅔ cup	liquid honey	150 mL

1. Combine apricots, lemon rind, water, lemon juice and ginger in a medium stainless steel or enamel saucepan. Bring to a boil over high heat, cover, reduce heat and boil gently for 35 minutes or until apricots are tender, stirring frequently.
2. Place apricot mixture in a food processor or blender and process until smooth; return to saucepan. Stir in honey. Return to a boil, reduce heat and boil gently, uncovered, until mixture is very thick, stirring frequently.
3. Ladle into sterilized jars and process as directed on page 16 (Shorter Time Processing Procedure).

Makes 2 cups (500 mL).

Microwave Lemon Curd

Lemon curd has long been a staple in many English households. It is fast gaining popularity in North America as an easy dessert served in tart shells, as a filling for meringue shells or to spread between layers of a cake or on scones. Making it in the microwave oven is *much* easier than making it in the traditional double boiler. Just be careful not to overcook it or it will separate.

2–3	lemons	2–3
¼ cup	butter	50 mL
¾ cup	granulated sugar	175 mL
2	eggs	2

1. Finely grate thin outer rind of lemons. Squeeze lemons. Measure ½ cup (125 mL) lemon juice into a 4-cup (1 L) microwavable container.
2. Stir in rind, butter and sugar. Microwave, uncovered, on High (100%) for 1½ to 2 minutes or until butter is melted and mixture is hot.
3. Beat eggs in a bowl. Gradually add hot lemon mixture to eggs, stirring constantly. Return mixture to the microwavable container and microwave, uncovered, on Medium (50%) for 1 to 2 minutes or just until thickened, stirring every 30 seconds. (Do not allow it to boil; mixture will thicken as it cools.) Let cool.
4. Pour curd into a tightly sealed container. Refrigerate up to 2 weeks or freeze for longer storage.

Makes 1⅔ cups (400 mL).

Variations:

Lime, Tangerine or Orange Curd: Use 1 lime, tangerine or orange in place of 1 lemon.

Spirited Curd: Stir 1 tbsp (15 mL) Amaretto or Grand Marnier into curd after cooking.

Tip: To get the maximum juice from citrus fruit, microwave fruit on High (100%) for 20 seconds before cutting.

Tip: Use Lemon Curd in Lady Fingers with Lemon Mousse (page 230).

Chapter Five

Light 'n' Low
Sugar Spreads

S OME JAMS are made the "lighter way," and for this reason go by the name "spreads." Government regulations require that a commercial jam exceed a specified percentage of sugar before it can be called a jam. Otherwise, it must be called a spread. We use this terminology for our home-made lighter jams. Since these lighter jams contain little sugar — an essential ingredient for making a true pectin gel — they are not as firm as other spreads.

Some people simply want a less-sweet spread. Others make this choice for a dietary reason, such as diabetes or weight loss. Whatever the reason, we think you will be pleased with our exciting spreads.

Recipes for our low-sugar spreads were developed using two sweeteners, aspartame (Equal) and granular low-calorie with sucralose (Splenda). The no-cook spreads tested well with both sweeteners. Sucralose, derived from sugar, retains its sweetness during cooking and gave us an excellent flavour with the cooked spreads. Aspartame tends to become bitter when cooked, so it's best added after cooking. Each recipe provides guidance for using each sweetener. Some of our recipes have a small amount of sugar added to the sweetener to enhance the fruit flavours. Unlike sugar, sweeteners do not affect formation of the gel. You can add any amount to suit your personal taste.

Several of our recipes for low-sugar spreads use added light pectin. This is a special type of pectin that allows a gel to form from fruit with little or no sugar. Without sugar present, these spreads may be bland and have a weaker gel than traditional spreads.

Cooked spreads should be processed in a boiling-water canner. Store the no-cook ones in a refrigerator for up to 3 weeks or freeze for longer storage. All our low-sugar spreads have been analyzed for Canadian Diabetes Association Food Choice Values to make them suitable for persons with diabetes. Food Choice Values and a nutrient analysis is given with each recipe.

Serving Suggestions

These spreads are used in much the same way as any jam (see page 17).

List of Recipes

Light Spiced Raspberry Spread

This fresh-tasting spread uses a minimum of sugar and no sweetener. Its sweetness comes mainly from apple juice concentrate and the natural sweetness of the fruit.

1	pkg (300 g) frozen unsweetened raspberries OR 3 cups (750 mL) fresh	1
1	tart apple, peeled, cored and chopped	1
½ cup	apple juice concentrate	125 mL
2 tbsp	granulated sugar	25 mL
1 tsp	grated lemon rind	5 mL
2 tsp	lemon juice	10 mL
Pinch	each ground ginger, cinnamon and nutmeg	Pinch
½ tsp	almond extract	2 mL

1. Mash raspberries in a medium stainless steel or enamel saucepan and measure; you should have 2 cups (500 mL). Add apple, apple juice, sugar, lemon rind and lemon juice. Bring to a boil over high heat, reduce heat and boil gently, uncovered, for 20 minutes or until mixture is thickened and spreadable, stirring frequently.
2. Stir in ginger, cinnamon and nutmeg; simmer for 3 minutes. Remove from heat and add almond extract.
3. Ladle into sterilized jars and process as directed on page 16 (Shorter Time Processing Procedure). Once opened, this spread is best kept in the refrigerator and used within 3 weeks.

Makes 2 cups (500 mL).

Each serving: 1/48 of recipe (2 tsp/10 mL)

1 ++ Extra

Nutritional Information
3 g carbohydrate, 0 g protein, 0 g fat,
1 g fibre, 1 mg sodium, 14 kcal (60 kJ)

Light Strawberry Pineapple Spread

This all-natural fruit spread uses ever-popular strawberries and pineapple juice concentrate for most of its sweetness.

5 cups	strawberries, washed and hulled	1.25 L
1	Granny Smith apple, peeled, cored and chopped	1
1 tsp	grated lemon rind	5 mL
½ cup	pineapple juice concentrate	125 mL
2 tbsp	granulated sugar	25 mL
2 tsp	lemon juice	10 mL
½ tsp	vanilla extract	2 mL

1. Mash strawberries in a medium stainless steel or enamel saucepan and measure; you should have about 3 cups (750 mL). Add apple, lemon rind, pineapple juice, sugar and lemon juice. Bring to a boil over high heat, reduce heat and boil gently, uncovered, for 20 minutes or until mixture is thickened and spreadable, stirring frequently.
2. Remove from heat and stir in vanilla extract.
3. Ladle into sterilized jars and process as directed on page 16 (Shorter Time Processing Procedure). Once opened, these spreads are best kept in the refrigerator and used within 3 weeks.

Makes 3½ cups (875 mL).

Each serving: 1/56 of recipe (1 tbsp/15 mL)

1 ++ Extra

Nutritional Information
3 g carbohydrate, 0 g protein, 0 g fat,
1 g fibre, 0 mg sodium, 12 kcal (50 kJ)

Light Citrus Strawberry Spread

The diced orange helps extend the strawberries, particularly useful if you are using more expensive out-of-season berries. The tangy spread is quite refreshing. If you find it too tart, add liquid sweetener to taste.

1	large orange	1
4 cups	strawberries, washed and hulled	1 L
1 tbsp	lemon juice	15 mL
2 tbsp	granulated sugar	25 mL
1	box (49 g) light fruit pectin crystals	1
1 cup	granular low-calorie sweetener with sucralose	250 mL

1. Grate 2 tsp (10 mL) rind from orange; place in a large stainless steel or enamel saucepan. Remove and discard remaining white rind from orange. Chop pulp and place in a 4-cup (1 L) measuring cup.
2. Mash strawberries; add to orange. You should have 3 cups (750 mL) fruit.
3. Combine fruit, lemon juice, sugar and pectin in saucepan; mix well. Bring to a boil over high heat, stirring constantly. Stir in sweetener, return to a boil and boil hard for 1 minute, stirring constantly.
4. Ladle into sterilized jars and process as directed on page 16 (Shorter Time Processing Procedure). Once opened, this spread is best kept in the refrigerator and used within 3 weeks.

Makes 3 cups (750 mL).

Each serving: 1/48 of recipe (1 tbsp/15 mL)

1 ++ Extra

Nutritional Information
3 g carbohydrate, 0 g protein, 0 g fat,
1 g fibre, 0 mg sodium, 11 kcal (50 kJ)

Light Three-Fruit Marmalade Spread

This low-sugar version of classic marmalade is quick and convenient to make any time of the year.

1	can (19 oz/540 mL) crushed pineapple	1
1 cup	finely chopped pink grapefruit sections	250 mL
1 tsp	grated orange rind	5 mL
1 cup	finely chopped orange sections	250 mL
2 tbsp	granulated sugar	25 mL
¼ cup	water	50 mL
1	box (49 g) light fruit pectin crystals	1
2 cups	granular low-calorie sweetener with sucralose	500 mL

1. Combine pineapple, grapefruit, orange rind, orange and sugar in a large stainless steel or enamel saucepan. Stir in water and pectin.
2. Bring to a full boil over high heat, stirring constantly. Stir in sweetener, return to a boil and boil hard for 1 minute, stirring constantly.
3. Ladle into sterilized jars and process as directed on page 16 (Shorter Time Processing Procedure). Once opened, this spread is best kept in the refrigerator and used within 3 weeks.

Makes 5 cups (1.25 L).

Each serving: 1/80 of recipe (1 tbsp/15 mL)

1 ++ Extra

Nutritional Information
3 g carbohydrate, 0 g protein, 0 g fat,
0 g fibre, 0 mg sodium, 12 kcal (50 kJ)

Light Microwave Peach Plum Butter

This spread has great flavour and colour with the thick consistency expected of a good fruit butter.

1 cup	finely chopped peeled peaches	250 mL
1 cup	finely chopped plums	250 mL
1 tbsp	water	15 mL
½ cup	granular low-calorie sweetener with sucralose	125 mL
½ tsp	ground cinnamon	2 mL
¼ tsp	ground ginger	1 mL

1. Combine peaches, plums and water in a 4-cup (1 L) microwavable container. Microwave, uncovered, on High (100%) for 5 minutes, stirring once. Microwave, uncovered, on High for 10 minutes or until mixture is very thick, stirring every 3 minutes.
2. Stir in sweetener, cinnamon and ginger.
3. Spoon spread into clean jars or plastic containers to within ½ inch (1 cm) of rim. Cover with tight-fitting lids. Label jars and refrigerate for up to 1 week or freeze for longer storage.

Makes 1 cup (250 mL).

Each serving: 1/24 of recipe (2 tsp/10 mL)

1 ++ Extra

Nutritional Information
3 g carbohydrate, 0 g protein, 0 g fat,
0 g fibre, 0 mg sodium, 11 kcal (50 kJ)

Light No-Cook Blueberry Plum Banana Spread

The distinctive flavours of these three fruits combine beautifully. Add to this the convenience of no-cook and freezer storage!

2 cups	fresh or frozen blueberries	500 mL
1 cup	bananas (about 2 bananas)	250 mL
1 cup	finely chopped purple plums (about 3 plums)	250 mL
½ cup	orange juice	125 mL
1 tbsp	lemon juice	15 mL
¾ cup	granular low-calorie sweetener with sucralose or aspartame	175 mL
2 tbsp	granulated sugar	25 mL
1	box (49 g) light fruit pectin crystals	1

1. Mash blueberries and bananas in a large bowl. Add plums, orange juice and lemon juice. Stir well to combine.
2. Combine sweetener, sugar and pectin. Gradually stir into fruit mixture. Let stand for 30 minutes, stirring occasionally.
3. Spoon spread into clean jars or plastic containers to within ½ inch (1 cm) of rim. Cover with tight-fitting lids. Label jars and refrigerate for up to 1 week or freeze for longer storage.

Makes 3½ cups (875 mL).

Each serving: 1/84 of recipe (2 tsp/10 mL)

1 ++ Extra

Nutritional Information
3 g carbohydrate, 0 g protein, 0 g fat,
0 g fibre, 0 mg sodium, 10 kcal (40 kJ)

Light No-Cook Kiwifruit Pineapple Spread

Lime, along with pineapple juice, gives an interesting background flavour to the distinctive taste of kiwifruit in this easy-to-make spread.

1 cup	finely chopped peeled kiwifruit	250 mL
1 cup	unsweetened pineapple juice	250 mL
1 tsp	grated lime rind	5 mL
2 tbsp	lime juice	25 mL
1 cup	granular low-calorie sweetener with sucralose or aspartame	250 mL
2 tbsp	granulated sugar	25 mL
1	box (49 g) light fruit pectin crystals	1

1. Combine kiwifruit, pineapple juice, lime rind and juice in a medium bowl; stir well.
2. Combine sweetener, sugar and pectin. Gradually stir into fruit. Let stand for 30 minutes, stirring occasionally.
3. Spoon spread into clean jars or plastic containers to within ½ inch (1 cm) of rim. Cover with tight-fitting lids. Label jars and refrigerate for up to 1 week or freeze for longer storage.

Makes 2¼ cups (550 mL).

Each serving: 1/54 of recipe (2 tsp/10 mL)

1 ++ Extra

Nutritional Information
3 g carbohydrate, 0 g protein, 0 g fat,
0 g fibre, 0 mg sodium, 11 kcal (50 kJ)

Tip: Remember, the best flavour comes from fruit that is ripe but not too soft. Kiwifruit when purchased are often too firm. Wait for them to soften.

Light No-Cook Strawberry Daiquiri Spread

A fruit spread version of the famous cocktail, its no-cook preparation gives it a really fresh taste. Also, it is possible to add liquid sweetener to suit your own taste.

4 cups	strawberries, washed and hulled	1 L
½ cup	unsweetened pineapple juice	125 mL
1 tsp	grated lime rind	5 mL
2 tbsp	lime juice	25 mL
2 tbsp	dark rum (optional)	25 mL
1½ cups	granular low-calorie sweetener with sucralose or aspartame	375 mL
2 tbsp	granulated sugar	25 mL
1	box (49 g) light fruit pectin crystals	1

1. Crush strawberries in a large bowl; you should have about 2 cups (500 mL). Add pineapple juice, lime rind, lime juice and rum (if using). Stir well.
2. Combine sweetener, sugar and pectin. Gradually stir into fruit mixture. Let stand for 30 minutes, stirring occasionally.
3. Spoon spread into clean jars or plastic containers to within ½ inch (1 cm) of rim. Cover with tight-fitting lids. Label and refrigerate for up to 1 week or freeze for longer storage.

Makes about 3 cups (750 mL).

Each serving: 1/48 of recipe (1 tbsp/15 mL)

1 ++ Extra

Nutritional Information
3 g carbohydrate, 0 g protein, 0 g fat,
1 g fibre, 0 mg sodium, 14 kcal (60 kJ)

For wonderful pear jams, marmalade and
conserve, see pages 38, 45, 64 and 74.

Condiments
of Choice

The tartness of plums, combined with amaretto,
makes for a great jam. (See page 28.)

Introduction to

Condiments of Choice

―――――――――――――――――――

W HETHER a piece of spiced apple or a peppery salsa, condiments truly are the "spice of life." These savoury, piquant, salty or spicy accompaniments to food are an easy way to make an otherwise ordinary meal special. Our recipes range from traditional Fast Favourite Garlic Dill Pickles (page 101) to more exotic ones like Papaya Mango Salsa (page 140). With an interesting assortment of condiments at hand, it's easy to add a welcome perkiness to simple meals.

We use a pickling process to make our condiments. The process preserves the vegetable and fruit ingredients with an acid, usually vinegar. Good pickling technique is essential to successful condiment making.

Name that Condiment

Pickle is a piece of vegetable or fruit that has been preserved with a salt and/or a vinegar mixture. Pickles may be either sweet or sour and may use herbs or spices to provide extra heat or flavour.

Relish is a pickle that has been chopped rather than left whole. Relishes can be sweet or sour, mild or hot.

Salsa is a Mexican word meaning "sauce." Salsas can be either cooked or fresh. The word has come to refer to a blend of vegetables and/or fruits with spices and herbs.

Chutney is a spicy condiment made from fruit, vinegar, sugar and spices. Its origin is in India where it had the Hindu name "chatni." Chutneys can be smooth or chunky and range in spiciness from mild to very hot.

Mustard is a sauce made from the seeds of the mustard plant. Its spiciness ranges from mild to hot depending on the variety of mustard seed used. "Prepared" mustards are mustards mixed with other ingredients.

Ketchup is a spicy mixture made from the juice of cooked vegetables and fruits. In North America it typically refers to one made from tomatoes.

Essential Pickling Ingredients

Vegetables and Fruits

Cucumber is the most common vegetable found in condiments. It is essential that cucumbers to be pickled are not waxed. The thin coat of wax on the skin of the typical smooth green cucumber available in stores throughout the year prevents pickling brine from penetrating the cucumber. English seedless cucumbers are unsuitable for pickling because of their very high water content. Pickling cucumber varieties are the smaller, squatter ones with a bumpy skin and are sometimes called Kirbys. Kirbys are generally used only for pickles but may also be eaten fresh.

Fruits as well as vegetables make interesting condiments. We have included recipes for traditional ones like Spiced Pickled Peaches (page 116), and we have recipes for relishes, salsas and chutneys using a variety of fruits.

Whether vegetables or fruit, the *best* results come from the *best* produce. Choose the freshest and highest quality you can find.

Vinegar

Vinegar is the essential ingredient in pickling. It gives the acidity necessary to preserve the produce as well as piquant flavours. White vinegar is most commonly used because it does not affect the colour of the condiment. Cider and malt vinegars do change the colour but are sometimes used for their interesting flavours. All our recipes require vinegars that have at least 5% acetic acid. Never use one with less. Check the label for the percentage of acetic acid. Avoid specialty vinegars, which often are lower in acid. Never reduce the amount of vinegar in a recipe. If you want a product that is less sour, add a pinch of sugar.

Salt

Salt affects both the flavour and the texture of the condiment. It is important to use only pickling salt. Table salt contains iodine, which can turn the condiment dark, as well as anti-caking agents, which can give a cloudy appearance.

Sugar and Spice

Sugar is generally added for flavour, but it also helps to keep the preserved condiment firm. Most recipes call for white granulated sugar, but brown sugar and maple syrup may also be used for their flavours.

To maintain clarity, spices added during cooking are usually used in their whole form tied in a small piece of cheesecloth or placed in a large tea ball. This makes it easy to remove them before processing. A method that gives a stronger spice flavour is to place them in the jar before packing the condiment ingredients. Ground spices are usually added to relishes and chutneys where clarity is not an issue. Purchase spices in small quantities to keep them fresh. Always store them in airtight containers away from heat, light and moisture. A spice rack over the stove may look attractive, but it is not a good place to keep spices fresh.

How To's of Pickling

The first step is often to put the vegetables, especially if they are cut in pieces, in a solution of pickling salt (called a brine) or to sprinkle them with dry pickling salt. This draws out their moisture, resulting in a firmer product. Soaking in brine requires more time but, in general, we find the product less salty than that produced by the dry salt method. The choice of method depends on the vegetable and the recipe. With either method, the soaked or sprinkled vegetable must be rinsed and drained to eliminate excess salt.

Preserving foods by pickling relies on an acid, vinegar, to discourage growth of bacteria. The easiest method is to pour a syrup made of vinegar combined with salt, sugar or spices specified in the recipe over the ingredients packed into jars. Another method, based on fermentation, is often used to make a kind of dill pickle and sauerkraut. We have chosen not to include fermentation in this book since it is a more complicated and less reliable process than the methods we have used.

Processing in a boiling-water canner to give a vacuum seal is the only way to go for homemade condiments. All our recipes provide information. Grandmother may not have bothered, but processing is absolutely essential to ensure the safety of your carefully made condiments. The heat produced by processing destroys the moulds that grow in high-acid foods and spoil the product. The airtight seal it creates prevents further contamination.

Essential Pickling Equipment

Condiments require little equipment that isn't found in the well-equipped kitchen. You need a large stainless steel or enamel *saucepan* for cooking, and a *boiling-water canner* and *jar lifter* for processing. Since most condiments are quite thick, a plastic *jar filler or wide-mouth funnel* is helpful for filling the jars. *Canning jars and lids* of any size can be used, but we like the pint (500 mL) jars for pickles and the half-pint (250 mL) jars for salsas, relishes and chutneys. The tiny half-cup (125 mL) jars are ideal for small amounts of savoury sauces and for gifts.

Procedure for Longer Time Boiling–Water Processing

On the following page is the step-by-step procedure for the processing of foods that require 10 minutes or more processing time. Use this procedure for condiments as directed in the recipes.

Longer Time Processing Procedure

(For food that requires 10 minutes or more processing time.)

If the recipe requires a preparation and cooking time longer than 20 minutes, begin preparation of the ingredients first. Then bring the water and jars in the canner to a boil while the prepared food is cooking. If the ingredients require a shorter preparation and cooking time, begin heating the canner before you start your recipe. The jars do not need to be sterile if the processing time is 10 minutes or longer, but they do need to be hot. Have a kettle with boiling water handy to top up the water level in the canner after you have put in the jars.

Steps for Perfect Processing

20 Minutes Before Processing
Partially fill a boiling-water canner with hot water. Place the number of clean mason jars into the canner needed to hold the quantity of finished food prepared in the recipe. Cover and bring water to a boil over high heat. This step generally requires 15 to 20 minutes, depending on the size of your canner.

5 Minutes Before Processing
Approximately 5 minutes before you are ready to fill the jars, place snap lids in hot or boiling water according to manufacturer's directions.

Filling Jars
Remove jars from canner and ladle or pack the food into hot jars to within ½ inch (1 cm) of top rim (head space). If the food is in large pieces, remove trapped air bubbles by sliding a rubber spatula between glass and food; readjust the head space to ½ inch (1 cm). Wipe jar rim to remove any stickiness. Centre snap lid on jar; apply screw band just until fingertip tight.

Processing Jars
1. Place jars in canner and adjust water level to cover jars by 1 to 2 inches (2.5 to 5 cm). Cover canner and return water to boil. Begin timing when water returns to a boil. Process for the exact time specified in the recipe.
2. Remove jars from canner to a surface covered with newspapers or with several layers of paper towels and cool for 24 hours. Check jar seals (sealed lids turn downward). Label jars with contents and date and store in a cool, dark place.

Chapter Six

Pickle Perfection

P OPULAR FOODS used to make pickles are beans, beets, carrots, cauliflower, cucumbers, onions (especially the small pearl ones), sweet and hot peppers and zucchini. Slightly less common, but in our opinion absolutely wonderful, are oranges, pumpkin, asparagus and watermelon rind. All these can be made in sweet, sour or hot versions and flavoured with such herbs as dill, mustard seeds, bay leaf, peppercorns — the possibilities are endless.

Remember, you need perfect produce for perfect pickles. This means the very freshest produce available. Too long between harvest and preparation can result in hollowed or shriveled pickles.

Several techniques help produce superior pickles.

- Salting the vegetables before making them into pickles to draw out some of the moisture to produce a firmer pickle.
- Removing a thin slice from the blossom end of cucumbers to ensure removal of an enzyme that can cause pickles to soften. (To be sure to get the blossom end, we have found it easiest to remove and discard a thin slice from each end.)
- Using fresh whole spices. For easy removal, tie them loosely in a piece of cheesecloth or place them in a large tea ball.
- Storing pickles a few weeks, preferably at least three, before nibbling. They become mellower and have a better flavour balance.
- Serving pickles well chilled.

We know that consumers have been making pickles for many years without processing them in a boiling-water canner. Although at first we were skeptical about the necessity of this step and the pickle quality that would result, we decided to follow current recommendations and processed our pickles in a boiling-water canner. The pickles were not adversely affected by being in the hot-water bath and we do feel more comfortable about their safety. In our experience, processing does not make pickles any less crisp provided the procedure given in each recipe is followed.

Serving Suggestions

Pickles are marvellous companions to such buffet-table offerings as cold or hot meats, cheeses, breads and salads. They pep up any sandwich or salad plate with colour and perky flavours.

List of Recipes

Fast Favourite Garlic Dill Pickles

Often called kosher-style dill pickles, these are quick to make. Use either small whole cucumbers or cut larger ones into quarters. For an additional interesting flavour, tuck a small dried hot red pepper into each jar.

8–10	small pickling cucumbers (about 3 lb/1.5 kg)	8–10
2 cups	white vinegar	500 mL
2 cups	water	500 mL
2 tbsp	pickling salt	25 mL
4	heads fresh dill OR 4 tsp (20 mL) dill seeds	4
4	small cloves garlic	4

1. Cut a thin slice from the ends of each cucumber.
2. Meanwhile, combine vinegar, water and salt in a saucepan and bring to a boil.
3. Remove hot jars from canner. Place 1 head fresh dill or 1 tsp (5 mL) dill seeds and 1 clove garlic into each jar; pack in cucumbers. Pour boiling vinegar mixture over cucumbers to within ½ inch (1 cm) of rim (head space). Process as directed on page 98 (Longer Time Processing Procedure) for 10 minutes for pint (500 mL) jars and 15 minutes for quart (1 L) jars.

Makes 4 pint (500 mL) jars.

Garlic may turn blue or green in the jar. Nothing to be alarmed about, it is only the effect of the acid on the natural pigments in the garlic.

Sweet Garlic Dills

These crisp garlicky pickles are adapted from a recipe shared with us by Jane and Jennifer, hosts of the television series *Put a Lid on It!*

12–16	small pickling cucumbers (about 4 lb/2 kg)	12–16
4	large cloves garlic	4
4	heads fresh dill OR 4 tsp (20 mL) dill seeds	4
½ tsp	celery seeds	2 mL
2 cups	white vinegar	500 mL
⅔ cup	water	150 mL
1 cup	granulated sugar	250 mL
3 tbsp	pickling salt	45 mL
Pinch	turmeric	Pinch

1. Cut a thin slice from the ends of each cucumber. Cut cucumbers lengthwise into quarters.
2. Remove hot jars from canner. Place 1 clove garlic, 1 head fresh dill or 1 tsp (5 mL) dill seeds and ⅛ tsp (0.5 mL) celery seeds into each jar; pack in cucumbers.
3. Meanwhile, combine vinegar, water, sugar, salt and turmeric in a small saucepan and bring to a boil. Pour boiling vinegar mixture over cucumbers to within ½ inch (1 cm) of rim (head space). Process for 10 minutes for pint (500 mL) jars and 15 minutes for quart (1 L) jars as directed on page 98 (Longer Time Processing Procedure).

Makes 4 pint (500 mL) jars.

Tip: If you can find small cucumbers approximately 6 inches (15 cm) long, they fit neatly into wide-mouth pint (500 mL) jars.

Many old recipes call for alum as a crisping agent for pickles. Alum is crystals of potassium aluminum sulfate and is no longer recommended for pickling. It can give a bitter flavour to pickles and may cause digestive upsets. Modern pickling procedures make it unnecessary.

Nine-Day Icicle Pickles

Icicle pickles have a long tradition. You probably remember your grandmother making them. Don't let the nine days deter you. The steps are simple — the rewards are well worth the effort.

2 quarts	pickling cucumbers, 4 to 6 inches (10 to 15 cm) long	2 L
4 cups	boiling water	1 L
½ cup	pickling salt	125 mL
2 cups	white vinegar	500 mL
3 cups	granulated sugar, divided	750 mL
1 tbsp	pickling spice	15 mL

Days 1–3

- Cut a thin slice from the ends of each cucumber. Cut cucumbers lengthwise into quarters. Cut each quarter crosswise in half. Place in a large non-reactive container. Combine boiling water and salt; pour over cucumbers. Place a weight such as a plate on top of cucumbers to keep them submerged. Stir once a day for 3 days.

Day 4

- Drain cucumbers and discard liquid. Cover cucumbers with fresh boiling water. Replace weight and let stand for 24 hours.

Day 5

- Drain cucumbers and discard liquid. Cover cucumbers with fresh boiling water. Replace weight and let stand for 24 hours.

Day 6

- Drain cucumbers and discard liquid.
- Prepare a brine: combine vinegar and 1½ cups (375 mL) sugar in a stainless steel or enamel saucepan. Place pickling spice in a large tea ball or tie in a piece of cheesecloth; add to saucepan. Bring to a full boil over high heat. Pour over cucumbers and let stand for 24 hours.

Day 7

- Drain brine into a large saucepan; add ½ cup (125 mL) sugar. Bring to a full boil over high heat. Pour over pickles and let stand for 24 hours.

Day 8

- Repeat Day 7, adding ½ cup (125 mL) sugar to the brine.

Day 9

- Partially fill a boiling-water canner with hot water. Place clean mason jars in canner. Begin to bring water to a boil over high heat. Cover and boil for at least 10 minutes to sterilize jars.
- Meanwhile, drain brine into large saucepan and add ½ cup (125 mL) sugar. Bring to a boil over high heat. Place snap lids in hot or boiling water according to manufacturer's directions.
- Remove jars from canner and pack pickles into jars. Pour hot brine over pickles to within ½ inch (1 cm) of rim (head space). Remove air bubbles by sliding a clean small wooden or plastic spatula between glass and pickles; readjust head space to ½ inch (1 cm). Wipe jar rim to remove any stickiness. Centre snap lid on jar; apply screw band just until fingertip tight. Place jars in canner and adjust water level to cover jars by 1–2 inches (2.5–5 cm). Cover canner and return water to a boil. Process for 5 minutes for pint (500 mL) and quart (1 L) jars.
- Remove hot jars from canner and cool for 24 hours. Check jar seals (sealed lids turn downward). Wipe jars, label and store in a cool, dark place.

Makes 3 or 4 pint (500 mL) jars.

The secret to crisp sweet pickles is adding the sugar gradually during the brining process.

Curry Pickle Slices

All curry lovers will be happy with this interesting flavour variation to a traditional pickle.

2 quarts	pickling cucumbers	2 L
4	small onions, sliced	4
1 tbsp	pickling salt	15 mL
2½ cups	cider vinegar	625 mL
1⅔ cups	granulated sugar	400 mL
1 tbsp	curry powder	15 mL
2 tsp	pickling spice	10 mL
1 tsp	each celery seeds and mustard seeds	5 mL

1. Cut a thin slice from the ends of each cucumber. Cut cucumbers into thick slices. Place with onions in a non-reactive container and sprinkle with salt. Let stand for 24 hours; drain and rinse twice.
2. Combine vinegar, sugar, curry powder, pickling spice, celery seeds and mustard seeds in a large stainless steel or enamel saucepan. Bring to a boil over high heat. Add vegetables and return just to a boil. Remove from heat.
3. Remove hot jars from canner. Remove vegetables from liquid with a slotted spoon; pack into jars. Pour liquid over vegetables to within ½ inch (1 cm) of rim (head space). Process for 10 minutes for pint (500 mL) jars and 15 minutes for quart (1 L) jars as directed on page 98 (Longer Time Processing Procedure).

Makes 4 pint (500 mL) jars.

Pickling Spice Blend
Pickling spice is a blend of various spices. Since we have had trouble finding it when it's not what stores call "the preserving season," here is a recipe.
Combine 2 tbsp (25 mL) each allspice berries, cardamom seeds, coriander seeds, whole cloves, mustard seeds and peppercorns. Add 2 bay leaves, crumbled; 2 cinnamon sticks, broken; 2 small pieces dried gingerroot, chopped, and 2 small dried red chilies, crushed (1–2 tsp/5–10 mL hot pepper flakes). Store in a tightly sealed container until ready to use.

Best Bread-and-Butter Pickles

Most of our mothers made their own versions of bread-and-butter pickles. If you yearn for this bit of nostalgia, try our favourite version.

4 lb	small pickling cucumbers	2 kg
4	small onions, thinly sliced	4
1	sweet green pepper, cut in thin strips	1
1	sweet red pepper, cut in thin strips	1
2 tbsp	pickling salt	25 mL
4 cups	cider vinegar	1 L
3 cups	granulated sugar	750 mL
2 tbsp	mustard seeds	25 mL
1 tsp	celery seeds	5 mL
½ tsp	turmeric	2 mL
¼ tsp	ground cloves	1 mL

1. Cut a thin slice from the ends of each cucumber. Slice cucumbers medium thick, about ³⁄₁₆ inch (4 mm). Place cucumbers, onions and peppers in a non-reactive container. Sprinkle with salt and let stand for 3 hours; drain. Rinse twice and drain thoroughly.
2. Combine vinegar, sugar, mustard seeds, celery seeds, turmeric and cloves in a large stainless steel or enamel saucepan. Bring to a boil over high heat. Add vegetables and return to a boil for 30 seconds or just until cucumbers are no longer bright green.
5. Remove hot jars from canner. Remove vegetables from liquid with a slotted spoon; pack into jars. Pour liquid over vegetables to within ½ inch (1 cm) of rim (head space). Process for 10 minutes for pint (500 mL) jars and 15 minutes for quart (1 L) jars as directed on page 98 (Longer Time Processing Procedure).

Makes 6 pint (500 mL) jars.

Tip: On a hot summer day you may want to keep the pickles cool during the salting. Adding ice cubes to the top of the cucumbers after sprinkling with salt is an easy way to do this.

Mixed Vegetable Mustard Pickles

This pickle is the original version of chow chow, Chinese pickled vegetables preserved in a sweet-tart syrup. This name has evolved to refer to any mixed vegetable pickle or relish made with mustard. Amazingly easy to make, it is a great way to convert a deluge of summer vegetables so they can be enjoyed later.

1 quart	pickling cucumbers (about 1¼ lb/625 g)	1 L
4 cups	cauliflower florets	1 L
	(about 1 small cauliflower)	
1 cup	peeled pearl onions	250 mL
½ cup	pickling salt	125 mL
6 cups	lukewarm water	1.5 L
3 cups	granulated sugar	750 mL
½ cup	all-purpose flour	125 mL
3 tbsp	dry mustard	45 mL
1 tbsp	celery seeds	15 mL
1½ tsp	turmeric	7 mL
3 cups	white vinegar	750 mL
½ cup	water	125 mL

1. Cut a thin slice from the ends of each cucumber. Cut cucumbers into thick chunks. Place cucumbers, cauliflower and onions in a large non-reactive container. Combine salt with lukewarm water, stirring until dissolved. Pour over vegetables and let stand for 24 hours. Rinse twice and drain thoroughly.

2. Combine sugar, flour, mustard, celery seeds and turmeric in a large saucepan; stir until well mixed. Whisk in vinegar and water. Bring to a boil over high heat, stirring constantly, until smooth and thickened. Add vegetables and return to a boil for 30 seconds.

3. Remove hot jars from canner. Remove vegetables from liquid with a slotted spoon; pack into jars. Pour liquid over vegetables to within ½ inch (1 cm) of rim (head space). Process for 10 minutes for pint (500 mL) jars and 15 minutes for quart (1 L) jars as directed on page 98 (Longer Time Processing Procedure).

Makes 5 pint (500 mL) jars.

Pickled Mushrooms

Pickled mushrooms are a wonderful side dish with salads and cold meats and to serve on an appetizer plate.

3 cups	button mushrooms, trimmed (about 1 lb/500 g)	750 mL
1 cup	white wine vinegar	250 mL
¼ cup	water	50 mL
3 tbsp	granulated sugar	45 mL
1½ tsp	each dried basil and tarragon	7 mL
½ tsp	dried thyme	2 mL
¼ tsp	pickling salt	1 mL
1	large clove garlic	1

1. Clean mushrooms. Combine vinegar, water, sugar, basil, tarragon, thyme and salt in a medium saucepan. Bring to a boil over high heat. Add mushrooms and garlic. Cover and boil gently for 5 minutes. Remove from heat; discard garlic.
2. Remove hot jars from canner. Remove mushrooms from liquid with a slotted spoon; pack into jars. Return liquid to a boil and pour over mushrooms to within ½ inch (1 cm) of rim (head space). Process for 10 minutes for half-pint (250 mL) jars as directed on page 98 (Longer Time Processing Procedure).

Makes 3 half-pint (250 mL) jars.

Tip: When cleaning mushrooms, never wash under running water — they will absorb water, resulting in a watery pickle. Instead, wipe the mushrooms with a dampened paper towel or use a mushroom brush.

Variation:

Replace basil and tarragon with 1 tbsp (15 mL) pickling spice or whole peppercorns.

Winter Salad Pickle

Make this pickle when all the fresh vegetables are plentiful. Keep lots on hand for a quick wintertime salad.

2 cups	cauliflower florets	500 mL
1 cup	peeled pearl onions, or larger onions cut into quarters	250 mL
1 cup	thickly sliced celery	250 mL
1 cup	sliced carrot	250 mL
1 cup	thickly sliced zucchini	250 mL
1 cup	yellow beans, trimmed and cut into 1-inch (2.5 cm) pieces	250 mL
2	medium sweet red peppers, cut into squares	2
3 cups	white wine vinegar OR Herb Vinegar (page 185)	750 mL
1½ cups	granulated sugar	375 mL
1⅓ cups	water	325 mL
2 tsp	pickling salt	10 mL
Pinch	paprika	Pinch

1. Combine cauliflower, onions, celery and carrot in a large bowl. Combine zucchini, beans and peppers in a separate bowl.
2. Combine vinegar, sugar, water, salt and paprika in a large stainless steel or enamel saucepan. Bring to a full boil over high heat. Add cauliflower, onions, celery and carrot and return just to a boil. Remove from heat and stir in zucchini, beans and peppers.
3. Remove hot jars from canner. Remove vegetables from liquid with a slotted spoon; pack into jars. Pour liquid over vegetables to within ½ inch (1 cm) of rim (head space). Process for 10 minutes for pint (500 mL) jars and 15 minutes for quart (1 L) jars as directed on page 98 (Longer Time Processing Procedure).

Makes 4 pint (500 mL) jars.

Variation:

Use any combination of vegetables for a total of 8 cups (2 L).

Easy Spiced Pickled Beets

Pickled beets have long been a favourite in our families. Tiny beets are the most attractive, but larger ones cut into pieces are just as delicious.

10–15	fresh beets	10–15
2 cups	granulated sugar	500 mL
2 cups	white vinegar	500 mL
⅓ cup	water	75 mL
16	whole cloves	16
8	whole allspice berries	8
2	cinnamon sticks	2
2 tsp	pickling salt	10 mL

1. Trim beets, leaving 1 inch (2.5 cm) of stem and tap root attached. Place beets in a large saucepan and cover with water. Bring to a boil over high heat, reduce heat, cover and simmer for 25 to 45 minutes or until tender. Drain and rinse under cold water. Remove skins and cut beets into large pieces.
2. Combine sugar, vinegar and water in a large saucepan. Bring to a boil over high heat, stirring occasionally.
3. Remove hot jars from canner and place 4 whole cloves, 2 allspice berries, ½ cinnamon stick and ½ tsp (2 mL) salt in each jar. Pack beet pieces into jars.
4. Pour hot liquid over beets to within ½ inch (1 cm) of rim (head space). Process for 30 minutes for pint (500 mL) jars and 35 minutes for quart (1 L) jars as directed on page 98 (Longer Time Processing Procedure).

Makes 4 pint (500 mL) jars.

Variation:

For an interesting variation, add ¼ tsp (1 mL) hot pepper flakes to each jar.

Pickled Jalapeño Peppers

People who enjoy lots of heat with their meals will keep these preserved jalapeños for the time an extra spice accompaniment to a meal is needed. Try Pickled Jalapeño Peppers in a variety of appetizers: Jalapeño Cheddar Canapés (page 214) or Picante Cream Cheese Dip (page 216). They also add lots of flavour to Jalapeño Quesadillas (page 218).

1 cup	cider vinegar	250 mL
¼ cup	water	50 mL
4 tsp	liquid honey	20 mL
2 tsp	pickling spice	10 mL
½ tsp	pickling salt	2 mL
2	cloves garlic, halved	2
½ lb	jalapeño peppers, seeded and thinly sliced	250 g

1. Combine vinegar, water, honey, pickling spice and salt in a small saucepan. Bring to a boil over high heat, remove from heat and let stand for 10 minutes.
2. Remove hot jars from canner. Place ½ clove garlic in each jar. Divide peppers between jars. Add ½ clove garlic.
3. Return pickling liquid to a boil. Pour over peppers to within ½ inch (1 cm) of rim (head space). Process for 10 minutes for half-pint (250 mL) jars and 15 minutes for pint (500 mL) jars as directed on page 98 (Longer Time Processing Procedure).

Makes 2 half-pint (250 mL) jars.

Tips:
- For extra eye-appeal, strips of either hot (cayenne, de Arbol and Mirasol are generally available) or sweet red peppers may be placed in jars with jalapeño slices before liquid is added.
- Other hot peppers may be processed in the same way. Scotch bonnet (habanero) are very hot and so a little will go a long way to add heat to meals. De Arbol and Mirasol are mid-heat peppers. See the Chile Pepper Heat Scale (page 133) and process them as for jalapeños.

Multi-Coloured Ginger Pickled Peppers

Allow these pickles to sit for several weeks for the full flavour to develop. Serve them with cold cuts or roasted meats and salads.

1	sweet green pepper, sliced lengthwise	1
1	sweet red pepper, sliced lengthwise	1
1	sweet yellow pepper, sliced lengthwise	1
2	jalapeño peppers, seeded and thinly sliced	2
1	2-inch (5 cm) piece gingerroot, peeled and thinly sliced	1
1½ cups	rice vinegar	375 mL
½ cup	water	125 mL
2 tbsp	granulated sugar	25 mL
1 tsp	pickling salt	5 mL

1. Place peppers and gingerroot in a shallow bowl. Combine vinegar, water, sugar and salt; stir well to dissolve. Pour over peppers. Cover and refrigerate overnight.
2. Drain peppers, reserving liquid. Remove hot jars from canner. Pack peppers into jars.
3. Bring drained liquid to a boil over high heat. Pour over peppers to within ½ inch (1 cm) of rim (head space). Process for 15 minutes for half-pint (250 mL) jars and 20 minutes for pint (500 mL) jars as directed on page 98 (Longer Time Processing Procedure).

Makes 3 half-pint (250 mL) jars.

Radish Refrigerator Pickles
Serve radishes as a colourful side-salad pickle. The colour from the radishes bleeds after assembling, so make them no longer than 2 hours before serving.
Combine ½ cup (125 mL) rice vinegar, 2 tbsp (25 mL) granulated sugar, 2 tsp (10 mL) finely chopped gingerroot and 1 tsp (5 mL) chopped fresh dill in a small bowl. Wash, trim and slice 2 bunches radishes; toss with dressing, cover and refrigerate up to 2 hours before serving. Makes 2 cups (500 mL).

Asparagus Pickles

The person who shared this recipe told us of a group who have a kitchen get-together every spring to make 150 lbs of asparagus into this pickle. They suggest placing wide-mouth pint (500 mL) canning jars on their side for easy filling with the asparagus spears.

2–2½ lb	asparagus spears	1–1.2 kg
4	small jalapeño peppers	4
4	cloves garlic	4
2 tsp	dill seeds	10 mL
3 cups	white vinegar	750 mL
¾ cup	water	175 mL
3 tbsp	granulated sugar	45 mL
1 tsp	pickling salt	5 mL
Pinch	coarsely ground black pepper	Pinch

1. Wash asparagus and cut each spear 4¼ inches (11 cm) long or long enough to fit a wide-mouth pint (500 mL) jar leaving ½ inch (1 cm) head space.
2. Remove hot jars from canner; place 1 jalapeño pepper, 1 clove garlic and ½ tsp (2 mL) dill seeds in each jar. Pack asparagus in jars with tips down.
3. Meanwhile, combine vinegar, water, sugar, salt and pepper in a medium saucepan and bring to a boil. Pour hot liquid over asparagus to within ½ inch (1 cm) of rim (head space). Process for 15 minutes for pint (500 mL) jars as directed on page 98 (Longer Time Processing Procedure).

Makes 4 pint (500 mL) jars.

Asparagus Ham Appetizer
Cut thinly sliced prosciutto or other ham into long rectangles. Wrap around asparagus pickle. Serve with Honey Mustard Sauce (page 163) or Roasted Red Pepper Mustard Sauce (page 162).

Carrot Zucchini Pickle Strips

Here is a neat pickle to make when zucchini are taking over the garden. Carrots and a hint of dill add interest.

1 lb	carrots, peeled and cut into short strips	500 g
1 lb	zucchini, cut into short strips	500 g
1 tsp	pickling salt	5 mL
1¾ cups	white vinegar	425 mL
⅔ cup	water	150 mL
⅓ cup	granulated sugar	75 mL
2 tbsp	chopped fresh dill	25 mL
2 tbsp	chopped fresh parsley	25 mL
½ tsp	each freshly ground pepper and dried thyme	2 mL

1. Cook carrots in boiling water for 2 minutes; drain and refresh in cold water. Combine carrots and zucchini; sprinkle with salt. Let stand for 4 hours; drain and rinse twice.
2. Mix vinegar, water, sugar, dill, parsley, pepper and thyme in a small saucepan. Bring to a boil, stirring until sugar has dissolved.
3. Remove hot jars from canner and pack vegetables into jars. Pour liquid over vegetables to within ½ inch (1 cm) of rim (head space). Process for 15 minutes for half-pint (250 mL) jars and 20 minutes for pint (500 mL) jars as directed on page 98 (Longer Time Processing Procedure).

Makes 5 half-pint (250 mL) jars.

Refrigerator Pickles
Add this easy refrigerator pickle to your condiment recipes. They are really delicious.
Thinly slice 1 unpeeled English cucumber. Layer ½ of cucumber with ½ thinly sliced small onion; repeat layers. In small saucepan combine ½ cup (125 mL) granulated sugar, ½ cup (125 mL) white vinegar, ¼ cup (50 mL) cider vinegar, pinch each salt, mustard seeds, celery seeds and ground turmeric. Bring to a boil for 1 minute. Pour over cucumber and onion; let cool. Cover and marinate in refrigerator for 4 days. These crisp pickles keep in the refrigerator for up to 1 month.

Pumpkin Pickles

This famous pickle comes from a friend of a friend. She suggests putting a light rather than a candle inside your Hallowe'en pumpkin so you can recycle it into pickles. Better still, paint a face on the pumpkin with magic markers.

1	large pumpkin	1
6 cups	granulated sugar	1.5 L
3 cups	white or cider vinegar	750 mL
1 tsp	whole cloves	5 mL
1	broken cinnamon stick	1
2	pieces crystallized ginger	2

1. Peel pumpkin, remove seeds and cut into 2-inch (5 cm) cubes.
2. Bring sugar and vinegar to a boil over high heat in a large stainless steel or enamel saucepan, stirring until sugar is dissolved. Tie cloves, cinnamon and ginger tightly in cheesecloth bag; add to sugar. Reduce heat and boil gently for 5 minutes.
3. Add pumpkin pieces and return to a boil. Reduce heat, cover, and boil gently for 25 minutes or until pumpkin is tender but pieces still hold their shape, stirring frequently. Discard spice bag.
4. Remove hot jars from canner. Remove pumpkin from liquid with a slotted spoon; pack tightly into jars. Pour liquid over pumpkin to within ½ inch (1 cm) of rim (head space). Process for 15 minutes for pint (500 mL) jars and 20 minutes for quart (1 L) jars as directed on page 98 (Longer Time Processing Procedure).

Makes 5 pint (500 mL) jars.

Tip:
Depending on the size of the pumpkin, you may need to make more of the syrup. Always use the same proportions called for in the recipe: 2 parts sugar to 1 part white vinegar.

Spiced Pickled Peaches

Both the peach and the pineapple version of this condiment is a wonderful accompaniment to many meals. Look for very small peaches. If you can find only large ones, cut them into quarters.

21–24	small peaches (about 4 lb/2 kg)	21–24
3½ cups	granulated sugar	875 mL
2 cups	white vinegar	500 mL
1 cup	water	250 mL
3	cinnamon sticks, 3 inches (8 cm) long	3
1 tbsp	whole cloves	15 mL
½ tsp	whole allspice	2 mL

1. Bring a saucepan of water to a boil over high heat. Dip peaches into boiling water for about 30 seconds or until skins will slip off easily. Peel peaches and place in a solution of 8 cups (2 L) water and 1 tsp (5 mL) lemon juice.
2. Bring sugar, vinegar and water to a boil over high heat in a large stainless steel or enamel saucepan, stirring until sugar is dissolved. Tie cinnamon, cloves and allspice in a cheesecloth bag; add to sugar. Reduce heat, cover and boil gently for 15 minutes.
3. Drain peaches and add to syrup. Return to a boil and boil gently for 5 minutes. Discard spice bag.
4. Remove hot jars from canner. Remove peaches from liquid with a slotted spoon; pack into jars. Pour liquid over peaches to within ½ inch (1 cm) of rim (head space). Process for 20 minutes for pint (500 mL) jars and 25 minutes for quart (1 L) jars as directed on page 98 (Longer Time Processing Procedure).

Makes 3 quart (1 L) jars, depending on size of peaches.

Variation:

Spiced Pineapple Pickles: Use 1 peeled and cored pineapple cut into chunks in place of the peaches. Increase water to 3 cups (750 mL) for syrup.

Watermelon Rind Pickles

Make this delightful pickle from a part of the watermelon that is often discarded. Leave a small amount of the pink flesh to give a bit of colour. Cut interesting shapes with canapé cutters. The secret of the crisp texture is to add the sugar gradually during the pickling process. The extra two days this requires is well worth the wait.

4 cups	peeled watermelon rind, cut into 1-inch (2.5 cm) cubes	1 L
¼ cup	pickling salt	50 mL
4 cups	water	1 L
2 cups	granulated sugar, divided	500 mL
1 cup	white vinegar	250 mL
1	lemon or lime, thinly sliced	1
1 tsp	whole cloves	5 mL
1 tsp	whole allspice	5 mL
2	cinnamon sticks, 3 inches (8 cm) long	2

Day 1

- Place watermelon rind in a large non-reactive bowl. Dissolve salt in the water and pour over rind. Let stand for 4 hours; drain and rinse twice.
- Place rind in a large stainless steel or enamel saucepan; cover with cold water. Bring to a boil over high heat, reduce heat, cover, and boil gently for 8 minutes, or just until tender. Drain; place in a large non-reactive bowl.
- Combine 1 cup (250 mL) sugar, vinegar, lemon slices, cloves, allspice and cinnamon in a saucepan. Bring to a boil, stirring until sugar is dissolved, and pour over rind. Place a weight such as a plate on top of rind to keep it submerged. Let stand for 24 hours.

Day 2

- Drain liquid from rind into a saucepan; add ½ cup (125 mL) sugar. Bring to a boil and pour over rind. Replace weight and let stand for 24 hours.

Day 3

- Partially fill a boiling-water canner with hot water. Place clean mason jars in canner, cover and begin to bring water to a boil over high heat.

- Drain liquid from rind into a saucepan; add ½ cup (125 mL) sugar. Bring to a boil. Add rind and return to a boil. Remove from heat.
- Meanwhile, place snap lids in hot or boiling water according to manufacturer's directions.
- Remove hot jars from canner. Remove cinnamon sticks from liquid and place one in each jar. Remove rind from liquid with a slotted spoon; pack into jars. Pour liquid over rind to within ½ inch (1 cm) of rim (head space). Remove air bubbles by sliding a small clean wooden or plastic spatula between glass and food; readjust head space to ½ inch (1 cm). Wipe jar rim to remove any stickiness. Centre snap lid on jar; apply screw band just until fingertip tight. Place jars in canner and adjust water level to cover jars by 1–2 inches (2.5–5 cm). Cover canner and return water to a boil. Process for 10 minutes for half-pint (250 mL) jars, 10 minutes for pint (500 mL) jars and 15 minutes for quart (1 L) jars.
- Remove jars from canner and cool for 24 hours. Check jar seals (sealed lids turn downward). Wipe jars, label and store in a cool, dark place.

Makes 2 pint (500 mL) jars.

Spiced Orange Slices

We were served these unusual orange pickle slices on recent trips to Australia. They came as an accompaniment to a cheese tray. Since they are a sweet-sour condiment, they are not a traditional main course pickle. Try them in Spiced Orange-Slice Salad (page 223).

4	large oranges	4
8 cups	hot water	2 L
1 tsp	salt	5 mL
1 cup	granulated sugar	250 mL
½ cup	lightly packed brown sugar	125 mL
½ cup	each cider vinegar and water	125 mL
¼ cup	corn syrup	50 mL
8	whole cloves	8
4	cardamom pods	4
4	cinnamon sticks, 3 inches (8 cm) long	4
½ tsp	peppercorns	2 mL

1. Combine whole unpeeled oranges, 8 cups (2 L) hot water and salt in a large saucepan. Bring to a boil, reduce heat, cover and simmer for 45 minutes or until fruit is tender. Drain oranges, discarding liquid, and cool.
2. Cut oranges in half crosswise and then into very thin slices.
3. Combine granulated sugar, brown sugar, vinegar, water, corn syrup, cloves, cardamom, cinnamon and peppercorns in a large saucepan. Stir over high heat until sugars have dissolved. Reduce heat and cook for 10 minutes. Add orange slices, cover and cook gently for 20 minutes. Remove from heat and let stand for 5 minutes. Remove and discard cardamom and cinnamon.
4. Remove hot jars from canner. Remove orange slices from liquid with a slotted spoon; pack into jars. Pour liquid and whole cloves over oranges to within ½ inch (1 cm) of rim (head space). Process for 10 minutes for half-pint (250 mL) jars and 15 minutes for pint (500 mL) jars as directed on page 98 (Longer Time Processing Procedure).

Makes 4 half-pint (250 mL) jars.

Chapter Seven

Ravishing Relishes

R ELISHES are similar to pickles but are chopped to a coarse texture. They are a marvellous tangy, spicy and often salty mixture of sweet or tart fruits and crisp vegetables. Relative to chutneys, relishes have sharper flavours and coarser textures. All relishes in this chapter are cooked and then processed. Storage allows their flavours to mellow and continue to develop. For best flavour, allow relishes to wait a few weeks before tasting.

Relish ingredients with a high water content, like zucchini, onions and cucumbers, should first be put in salt water or layered with pickling salt to draw off some of their moisture. They are then drained and rinsed well before being cooked. Relish vegetables and fruits are cooked only until they are just tender to maintain their crisp texture. Most relish ingredients are available year round.

Serving Suggestions

Serve relishes with hot or cold meats as well as with fish and poultry such as duck and goose. Cottage or ricotta cheese, plain yogurts and salad vinaigrettes benefit from the addition of a relish. We enjoy Zucchini Garden Pepper Relish (page 123) added to mayonnaise to make an easy Thousand Island–style salad dressing. Relishes are also an essential ingredient in egg, chicken and tuna salad sandwich fillings.

The Antipasto Relish (page 126) is one of our favourites. The recipe contains directions for converting the relish to an antipasto by adding anchovies, tuna, olives and mushrooms at serving time. This is a very quick and convenient way to make an antipasto and eliminates the preserving problems associated with the low-acid nature of this condiment.

List of Recipes

Sweet Onion and Fennel Relish

Fennel's sweet, delicate hint of licorice is evident in this unique relish. It complements cold meats.

1	large sweet onion, such as Spanish or Vidalia (about 8 oz/250 g)	1
1	fennel bulb (about 10 oz/275 g)	1
1	sweet red pepper, sliced into thin strips	1
2½ tsp	pickling salt, divided	12 mL
1½ cups	white wine vinegar	375 mL
½ cup	water	125 mL
¼ cup	granulated sugar	50 mL
2	bay leaves	2
8	black peppercorns	8

1. Slice onion in half lengthwise, then in very thin slices crosswise to form half circles. Cut fennel bulb in half lengthwise and remove core; thinly slice crosswise to form half circles. Place onion, fennel and pepper in a non-reactive bowl and sprinkle with 2 tsp (10 mL) salt. Toss and let stand for 4 hours. Rinse twice and drain thoroughly.

2. Combine vinegar, water, sugar and ½ tsp (2 mL) salt in a large stainless steel or enamel saucepan. Bring to a boil over high heat. Add vegetables and return just to a boil, stirring constantly. Remove from heat.

3. Remove vegetables from liquid with a slotted spoon and pack into hot jars. Pour liquid over vegetables to within ½ inch (1 cm) of rim (head space). Add bay leaves and peppercorns. Process for 10 minutes for half-pint (250 mL) jars and pint (500 mL) jars as directed on page 98 (Longer Time Processing Procedure).

Makes 4 cups (1 L).

Zucchini Garden Pepper Relish

Zucchini lends a lightness and freshness to traditional pepper relish. Zucchini growers will love to have another such tasty use for their fast-growing vegetable.

4	medium zucchini (about 1¼ lb/625 g), finely chopped	4
2	medium onions, finely chopped	2
½	sweet red pepper, finely chopped	½
½	sweet green pepper, finely chopped	½
2 tbsp	pickling salt	25 mL
1¼ cups	granulated sugar	300 mL
¾ cup	cider vinegar	175 mL
1 tsp	each dry mustard and celery seeds	5 mL
½ tsp	each hot pepper flakes and turmeric	2 mL
1 tbsp	water	15 mL
2 tsp	cornstarch	10 mL

1. Toss together zucchini, onions and red and green peppers in a large non-reactive bowl. Sprinkle with salt and stir well. Let stand for 1 hour, stirring occasionally.
2. Drain vegetables in a sieve and rinse; drain again, pressing out excess moisture.
3. Combine drained vegetables, sugar, vinegar, mustard, celery seeds, hot pepper flakes and turmeric in large stainless steel or enamel saucepan. Bring to a boil over high heat, reduce heat and boil gently, uncovered, for 15 minutes or until vegetables are tender.
4. Blend water and cornstarch; stir into vegetables. Cook for 5 minutes or until liquid clears and thickens, stirring often.
5. Remove hot jars from canner and ladle relish into jars to within ½ inch (1 cm) of rim (head space). Process for 10 minutes for half-pint (250 mL) jars and 15 minutes for pint (500 mL) jars as directed on page 98 (Longer Time Processing Procedure).

Makes 4 cups (1 L).

Sunshine Mustard Relish

Variations of this relish have been known for several generations in Canada by such royal names as Lady Rose, Lady Ross and Lady Ashburnham. We adapted this recipe from one given us by the two sisters who appear in the television series *Put a Lid on It!*

3 cups	chopped cucumbers (about 3 medium)	750 mL
3 cups	chopped cauliflower (about ½ head)	750 mL
1 cup	chopped onion	250 mL
2 cups	water	500 mL
3 tbsp	pickling salt	45 mL
2	apples, peeled, cored and chopped	2
1	sweet red pepper, chopped	1
3 cups	granulated sugar	750 mL
2 tbsp	dry mustard	25 mL
1 tbsp	each mustard seeds and celery seeds	15 mL
1 tsp	turmeric	5 mL
3 cups	white vinegar, divided	750 mL
⅓ cup	all-purpose flour	75 mL

1. Combine cucumber, cauliflower and onion in a large bowl. Stir water and salt together and pour over vegetables. Let stand for 12 hours. Drain vegetables and rinse twice. Add apples and red pepper.
2. Combine sugar, dry mustard, mustard seeds, celery seeds and turmeric in a very large stainless steel or enamel saucepan. Stir in 2½ cups (625 mL) vinegar. Bring to a boil over high heat. Add vegetables and return to a boil; reduce heat and boil gently for 25 minutes, stirring occasionally.
3. Whisk together remaining ½ cup (125 mL) vinegar and flour. Stir into vegetables. Bring to a boil over high heat and cook for 1 minute.
4. Remove hot jars from canner and ladle relish into jars to within ½ inch (1 cm) of rim (head space). Process for 10 minutes for half-pint (250 mL) jars and 15 minutes for pint (500 mL) jars as directed on page 98 (Longer Time Processing Procedure).

Makes 10 cups (2.5 L) relish.

Tip: If your cucumbers are large, remove seeds before using.

Fresh Sweet Pepper and Peach Salsa is a delicious complement to chicken and pork dishes. (See page 144.)

Beet Relish with Horseradish

Horseradish gives this savoury beet relish extra zest to enhance any meat dish.

5	medium beets	5
1	large onion, finely chopped	1
2	sweet red peppers, finely chopped	2
1 cup	white vinegar	250 mL
½ cup	granulated sugar	125 mL
1 tsp	pickling salt	5 mL
⅔ cup	grated fresh horseradish	150 mL

1. Cook beets in boiling water until tender, about 20 minutes. Drain beets, remove skins and chop finely. There should be about 2 cups (500 mL). Mix beets with onions and peppers.
2. Combine vinegar, sugar, salt and horseradish in a large stainless steel or enamel saucepan. Bring to a boil over high heat. Add vegetables. Return to a boil, reduce heat and simmer, uncovered, for 20 minutes, stirring occasionally.
3. Remove hot jars from canner and ladle relish into jars to within ½ inch (1 cm) of rim (head space). Process for 15 minutes for half-pint (250 mL) jars and 20 minutes for pint (500 mL) jars as directed on page 98 (Longer Time Processing Procedure).

Makes 3½ cups (875 mL).

Tips:
- A can (14 oz/398 mL) of beets may be used in place of fresh for this relish.
- Commercially prepared horseradish may be substituted for the fresh, but double the amount.

Three kinds of tomatoes plus a range of herbs, spices and red wine make for a fabulous tomato sauce. (See page 175.)

Antipasto Relish

This recipe becomes antipasto when you add tuna, olives, mushrooms and anchovies.

2	large cloves garlic, minced	2
¾ cup	white vinegar	175 mL
½ cup	each water and tomato sauce	125 mL
¼ cup	granulated sugar	50 mL
2 tsp	pickling spice	10 mL
6	peppercorns	6
3	bay leaves	3
1 tsp	dried oregano	5 mL
1 cup	each small broccoli and cauliflower florets	250 mL
¾ cup	bottled small pickled onions, drained	175 mL
½	each large sweet red, yellow and green pepper, diced	½
2	carrots, peeled and thinly sliced	2
1	jalapeño pepper, seeded and finely chopped	1

1. Combine garlic, vinegar, water, tomato sauce and sugar in large stainless steel or enamel saucepan. Tie pickling spice, peppercorns and bay leaves in a cheesecloth bag. Add spice bag and oregano to saucepan. Bring to a boil over high heat, reduce heat and stir until sugar is dissolved. Add broccoli and cauliflower florets, pickled onions, sweet peppers, carrots and jalapeño pepper. Return to a boil, reduce heat and boil gently for 1 hour or until mixture has thickened. Discard spice bag.
2. Remove hot jars from canner and ladle relish into jars to within ½ inch (1 cm) of rim (head space). Process for 10 minutes for half-pint (250 mL) jars and 15 minutes for pint (500 mL) jars as directed on page 98 (Longer Time Processing Procedure).

Makes 3½ cups (875 mL).

Antipasto Appetizer
To 1 cup (250 mL) Antipasto Relish, add 1 can (6.5 oz/184 g) flaked and drained tuna, ½ cup (125 mL) each black and green olives, ½ cup (125 mL) canned sliced mushrooms and 2 flat anchovy fillets, minced. Keep refrigerated for up to 1 week or freeze for longer storage. Makes about 3 cups (750 mL).

Brinjal Pickle Relish

Ellie first tasted this relish in Australia and immediately set about finding a way to duplicate it. She found the secret of the relish's flavour in an old cookbook given to her by a visitor from Sri Lanka — sautéing the diced eggplant in "½ bottle of oil" before mixing it with "spiced pickle."

⅓ cup	vegetable oil	75 mL
1	eggplant, cut into ½-inch (1 cm) cubes	1
2–3	hot red chile peppers, seeded and finely chopped	2–3
3	large cloves garlic, finely chopped	3
¾ cup	white vinegar	175 mL
4 tsp	chili powder	20 mL
1 tbsp	whole fenugreek	15 mL
1 tsp	ground coriander	5 mL
½ tsp	dry mustard	2 mL
¼ tsp	each turmeric and salt	1 mL
½ cup	brown sugar	125 mL

1. Heat oil over medium heat in a large nonstick skillet. Add eggplant and fry gently for about 10 minutes. (At first the oil is completely absorbed, but then is gradually released as the eggplant becomes fairly firm.) Stir in chile peppers and garlic; cook for 3 minutes.
2. Stir in vinegar, chili powder, fenugreek, coriander, mustard, turmeric and salt. Bring to a boil, reduce heat and boil gently for about 10 minutes. Add sugar and cook for 2 minutes.
3. Remove hot jars from canner and ladle relish into jars to within ½ inch (1 cm) of rim (head space). Process for 15 minutes for half-pint (250 mL) jars and 20 minutes for pint (500 mL) jars as directed on page 98 (Longer Time Processing Procedure).

Makes 2 cups (500 mL).

Tip: Be sure to label your jars after processing and include the date. It's amazing how easy it is to forget what is inside and the year it was made.

Country Corn Relish

This colourful relish is excellent served with baked ham or roast pork, turkey and chicken. We also like it with cold meats and barbecued burgers. It also becomes a soup (page 220).

1¼ cups	cider vinegar	300 mL
2 cups	finely chopped celery (4 large stalks)	500 mL
1	large sweet red pepper, finely chopped	1
½	large sweet green pepper, finely chopped	½
1	large onion, finely chopped	1
½ cup	granulated sugar	125 mL
1 tsp	each celery seeds and pickling salt	5 mL
4 tsp	all-purpose flour	20 mL
2 tsp	dry mustard	10 mL
½ tsp	turmeric	2 mL
4 cups	fresh or frozen corn kernels	1 L
2 tbsp	chopped fresh parsley	25 mL

1. Combine vinegar, celery, red and green peppers, onion, sugar, celery seeds and salt in a large stainless steel or enamel saucepan. Bring to a boil over high heat and cook for 5 minutes.
2. Stir together flour, mustard, turmeric and 3 tbsp (45 mL) water. Whisk some of the hot pickling liquid into flour, then return to saucepan; boil gently, stirring until mixture is slightly thickened. Stir in corn and return to a boil.
3. Remove hot jars from canner. Stir parsley into relish and ladle relish into jars to within ½ inch (1 cm) of rim (head space). Process for 15 minutes for pint (500 mL) jars and half-pint (250 mL) jars as directed on page 98 (Longer Time Processing Procedure).

Makes 6 cups (1.5 L).

Green and Red Pepper Relish

We think this pepper relish is vastly superior to any commercial one. In fact, when sweet green and red peppers are in season, we always try to make our winter's supply.

4	sweet green peppers, chopped	4
4	sweet red peppers, chopped	4
4	medium onions, finely chopped	4
1 cup	white vinegar, divided	250 mL
1 cup	granulated sugar	250 mL
1 tsp	pickling salt	5 mL

1. Combine peppers, onions and ¾ cup (175 mL) boiling water in a large stainless steel or enamel saucepan. Cover and let stand for 5 minutes.
2. Drain vegetables and return to saucepan. Stir in ⅓ cup (75 mL) each vinegar and water. Bring to a boil, cover and reduce heat; simmer for 5 minutes.
3. Drain vegetables and return to saucepan. Heat remaining ⅔ cup (150 mL) vinegar, sugar and salt in a 2-cup (500 mL) microwavable container on High (100%) until sugar is dissolved. Add to vegetables and return mixture to a boil. Boil gently, uncovered, for 25 minutes or until liquid is reduced and vegetables are tender-crisp.
4. Remove hot jars from canner and ladle relish into jars to within ½ inch (1 cm) of rim (head space). Process for 10 minutes for half-pint (250 mL) jars and pint (500 mL) jars as directed on page 98 (Longer Time Processing Procedure).

Makes 4 cups (1 L).

Tip: Store empty mason jars with the used lids in place to keep them clean for the next use.

Cranberry Rum Relish

Take this tangy cranberry relish to your next turkey dinner. We love it as an accompaniment to grilled chicken breasts and it is also splendid with pâté and crackers.

⅓ cup	dark rum	75 mL
¼ cup	finely chopped shallots	50 mL
	Grated rind of 1 orange	
3 cups	fresh or frozen cranberries	750 mL
1 cup	granulated sugar	250 mL
½ tsp	freshly ground pepper	2 mL

1. Combine rum, shallots and orange rind in a medium saucepan. Bring to a boil over high heat, reduce heat and simmer for a few minutes until rum has reduced and mixture is a syrupy glaze.
2. Add cranberries and sugar. Stirring constantly, cook until cranberries pop and sugar is dissolved. Remove from heat and stir in pepper.
3. Remove hot jars from canner and ladle relish into jars to within ½ inch (1 cm) of rim (head space). Process for 10 minutes for half-pint (250 mL) jars and 15 minutes for pint (500 mL) jars as directed on page 98 (Longer Time Processing Procedure).

Makes 2 cups (500 mL).

Kiwifruit Relish
Kiwifruit makes an interesting refrigerated relish to serve with cold meats.
In a small saucepan, combine 1 cup (250 mL) each granulated sugar and white vinegar, ¼ cup (50 mL) water and 1 tsp (5 mL) each mustard seeds and peppercorns. Bring to a boil over high heat; remove from heat and cool. Meanwhile, place ½ cup (125 mL) diced shallots and 1 sweet red pepper, cubed, in a bowl. Cover with boiling water and let stand for 2 minutes; drain, refresh in cold water and pat dry. Peel and slice 5 firm kiwifruit. Combine kiwifruit, shallot mixture and vinegar mixture. Place in a clean jar and refrigerate for at least 5 days before tasting. Makes 2½ cups (625 mL).

Apricot Red Pepper Relish

This gourmet relish is superb both as a topping to cream cheese spread on crackers and as a zippy complement to roasts.

1½ cups	cider vinegar	375 mL
1 cup	diced sweet red pepper	250 mL
¼ cup	drained Pickled Jalapeño Peppers (page 111)	50 mL
1¼ cups	chopped dried apricot halves	300 mL
3½ cups	granulated sugar	875 mL
1	pouch (85 mL) liquid fruit pectin	1

1. Combine vinegar, red pepper and jalapeño pepper in blender or food processor. Process with on/off motion until finely chopped but not puréed. Transfer to large saucepan.
2. Add apricots and sugar. Bring to a boil, reduce heat and cook for 5 minutes. Remove from heat and stir in pectin.
3. Ladle relish into sterilized jars to within ½ inch (1 cm) of rim (head space). Process as directed on page 16 (Shorter Time Processing Procedure).

Makes 5 cups (1.25 L).

Variation:

For extra spirit, add 2 tbsp (25 mL) port to mixture during cooking.

Chapter Eight

Salsa Sensations

S ALSAS ARE DEFINED as a Mexican sauce that may be either cooked or fresh. Marketing statistics show salsas to be one of the fastest-growing segments in the supermarket sauce category. But have you noticed their price? Our book gives you the alternative of making your own. We provide a wide selection of salsa recipes, ranging from fresh no-cook and refrigerated, as in Pico de Gallo Salsa (page 142), to cooked and processed year round, as in Winter Salsa (page 137) and Garden Patch Salsa (page 138). You will find them every bit as good as any you can buy (if not better). *And* you'll find many exciting new ones that will never be found on a grocer's shelf.

Fresh uncooked salsas should be kept refrigerated and consumed quickly, as you would with any other fresh food. Most salsas are made from assorted vegetables, but fruit salsas are gaining popularity. So we have added several—Cranberry (page 139) and Papaya Mango (page 140) are examples. We also offer different "heat" levels to satisfy everyone's taste.

Speaking of heat levels, see our chart on the following page. It describes heat levels found in various peppers as reported by the Texas Agricultural Experiment Station at New Mexico State University. We hope it will help guide you when choosing from the ever-widening varieties available in stores. Chile peppers are interchangeable in most recipes, so experiment with their differing flavours and heat levels. Keep in mind that the chart's rating is specific to the particular pepper pod that was tested. Growing conditions, seasonal variability and cross-breeding make for great variations in heat levels.

BEWARE! It is important to wear rubber gloves when handling all hot chile peppers. Furthermore, wash your hands thoroughly with hot water and soap before touching your face, eyes or mouth.

Serving Suggestions

Salsas add flavour without adding fat. Such Mexican foods as fajitas, enchiladas and tacos would not be the same without them. Try a salsa as a base for a Mexican pizza or combined with light sour cream or yogurt for a great veggie dip. Be adventurous and mix a salsa with pasta for a speedy supper. And adding a salsa to ground meat before shaping it into patties provides fantastic flavour plus moisture to the humble hamburger.

Chile Pepper Heat Scale

Rating	Varieties
10 (hottest)	Habanero, Scotch Bonnet
9	Santaka, Chiltepin, Thai
8	Aji, Rocoto, Piquin, Cayenne, Tabasco
7	de Arbol
6	Yellow Hot Wax, Serrano
5	Jalapeño, Mirasol
4	Sandia, Cascabel
3	Ancho, Pasilla, Española
2	NuMex Big Jim
1	Mexi-Bell, Cherry
0	Sweet Bell, Pimiento, Sweet Banana

List of Recipes

Your Basic Chunky Tomato Salsa

Make this basic salsa in the fall when ingredients are at their freshest. We believe its many variations will suit your family's preferences.

4	medium tomatoes, peeled and chopped (about 1 lb/500 g)	4
1	medium onion, finely chopped	1
½	sweet green pepper, chopped	½
1–4	jalapeño peppers, halved, seeded and chopped	1–4
3	cloves garlic, minced	3
½ cup	tomato sauce	125 mL
½ cup	red wine vinegar	125 mL
½ cup	chopped fresh parsley	125 mL
2 tsp	granulated sugar	10 mL
½ tsp	pickling salt	2 mL
½ tsp	ground cumin	2 mL

1. Combine tomatoes, onion, green pepper, jalapeño peppers, garlic, tomato sauce, vinegar, parsley, sugar, salt and cumin in a large stainless steel or enamel saucepan. Bring to a boil over high heat, reduce heat and boil gently, uncovered, for 25 minutes or until desired consistency, stirring frequently.
2. Remove hot jars from canner and ladle salsa into jars to within ½ inch (1 cm) of rim (head space). Process for 20 minutes for half-pint (250 mL) jars and for pint (500 mL) jars as directed on page 98 (Longer Time Processing Procedure).

Makes 2½ cups (625 mL).

Variations:

Mild Salsa: Use 1 jalapeño pepper
Medium Salsa: Use 2 or 3 jalapeños
Hot Salsa: We guess the sky's the limit. Choose 1 Scotch bonnet, habanero or Jamaican chile pepper to replace the jalapeño peppers.
Basil or Thyme Tomato Salsa: Replace parsley with fresh basil or thyme.
Sherry Vinegar Salsa: Replace red wine vinegar with Sherry Vinegar (page 188).

Yagottabecrazy Salsa

You can be assured that any recipe calling for Scotch bonnet peppers will be HOT! No salsa chapter would be complete without one *really hot* one. Here it is. Our friend Kevin Myers, producer of the television series *Put a Lid on It!*, gave us this great recipe.

	Juice and grated rind of 1 lime	
2	cloves garlic	2
2	green onions	2
1–2	Scotch bonnet peppers, seeded	1–2
6	plum tomatoes, coarsely chopped	6
¼ cup	packed fresh cilantro leaves	50 mL
¼ cup	red wine vinegar	50 mL
¼ cup	water	50 mL
1 tsp	each pickling salt and granulated sugar	5 mL

1. Place lime juice, rind, garlic, onions and chile peppers in a food processor; chop finely. Add tomatoes and cilantro; process until fairly smooth. Transfer to a medium stainless steel or enamel saucepan.
2. Add vinegar, water, salt and sugar to tomato mixture. Bring to a boil, reduce heat and cook, uncovered, for 15 minutes or until mixture starts to thicken.
3. Remove hot jars from canner and ladle salsa into jars to within ½ inch (1 cm) of rim (head space). Process for 20 minutes for half-pint (250 mL) jars and for pint (500 mL) jars as directed on page 98 (Longer Time Processing Procedure).

Makes 1½ cups (375 mL).

Tip: Even with the seeds removed, these chiles are HOT. If you love heat, this one is for you. But if you wish to keep this heat under control, put the chiles in whole during cooking and remove them before canning. And don't forget when handling hot chiles, use rubber gloves.

Winter Salsa

How many times have you run out of *the most popular* condiment in the house, salsa —and the ingredients are out of season. Not to worry. This quickly prepared salsa made from a can of tomatoes, plus spices and a few other vegetables, fills the bill.

1	medium onion	1
2	large cloves garlic	2
½	sweet red or green pepper	½
1	can (19 oz/540 mL) tomatoes	1
2 tbsp	cider vinegar	25 mL
1 tsp	each granulated sugar, chili powder and dried oregano	5 mL
½ tsp	each salt and cumin	2 mL
¼ tsp	each pepper and paprika	1 mL
¼ cup	chopped fresh cilantro	50 mL

1. Finely chop onion, garlic and red pepper in a food processor or by hand. Place in a medium stainless steel or enamel saucepan.
2. Drain tomatoes in a sieve set over the saucepan. Chop tomatoes and add to saucepan. Stir in vinegar, sugar, chili powder, oregano, salt, cumin, pepper and paprika. Bring to a boil over high heat. Reduce heat and boil gently, uncovered, for 30 minutes or until mixture is thickened, stirring occasionally. Stir in cilantro 5 minutes before ladling into jars.
3. Remove hot jars from canner and ladle salsa into jars to within ½ inch (1 cm) of rim (head space). Process for 10 minutes for half-pint (250 mL) jars and 15 minutes for pint (500 mL) jars as directed on page 98 (Longer Time Processing Procedure).

Makes 3 cups (750 mL).

Salsa Sense
- Add salsa to ground chicken or turkey when making burgers. Extra can be served with the cooked burgers.
- Top cooked salmon fillets with any salsa before either cooking or serving.
- Add salsa to light sour cream or plain yogurt for a low-fat dip.
- Stir some salsa into your next batch of ground beef.
- Make your own taco filling by adding lots of salsas to the cooked meat.

Garden Patch Salsa

This mixed vegetable salsa is an excellent accompaniment to burgers and cheese nachos or as a low-fat topping for a baked potato. Try it as a zesty dip.

6	tomatoes, peeled and diced	6
4	jalapeño peppers, seeded and minced	4
2	cloves garlic, minced	2
1 cup	chopped onion	250 mL
1 cup	each shredded carrot and shredded zucchini	250 mL
½ cup	each chopped sweet green pepper and chopped sweet yellow pepper	125 mL
½ cup	chopped Italian (flat-leaf) parsley	125 mL
½ cup	white vinegar	125 mL
⅓ cup	tomato paste	75 mL
¼ cup	chopped fresh oregano OR 1 tbsp (15 mL) dried	50 mL
½ tsp	pickling salt	2 mL

1. Place all ingredients in a large stainless steel or enamel saucepan. Bring to a boil over high heat, reduce heat and simmer, uncovered, for 30 minutes or until thickened.
2. Remove hot jars from canner and ladle salsa into jars to within ½ inch (1 cm) of rim (head space). Process for 20 minutes for half-pint (250 mL) and pint (500 mL) jars as directed on page 98 (Longer Time Processing Procedure).

Makes 5½ cups (1.375 L).

Mexican Vegetable Salsa Pizza
Place a 12-inch (30 cm) pizza crust or flatbread round on a baking sheet and spread with your favourite tomato salsa. Spoon 1 drained can (14 oz/398 mL) black beans or lentils over top and sprinkle with 2 finely chopped jalapeño peppers and ½ tsp (2 mL) cumin. Bake at 450°F (230°C) for 10 minutes. Sprinkle with 2 cups (500 mL) shredded mozzarella cheese and bake 10 minutes or until cheese is melted and golden.

Cranberry Salsa

A marriage of Mexican and gringo flavours provides a spicy twist to traditional cranberry sauce. Use wherever you would use cranberry sauce.

1½ cups	fresh or frozen cranberries	375 mL
¾ cup	frozen orange juice concentrate, thawed	175 mL
1	sweet yellow or red pepper, finely chopped	1
1	long yellow hot wax pepper, seeded and finely chopped	1
1 cup	finely chopped onion	250 mL
1	clove garlic, minced	1
¼ cup	red wine vinegar	50 mL
¼ cup	packed brown sugar	50 mL
½ tsp	each ground cumin and pickling salt	2 mL
½ cup	loosely packed fresh cilantro, coarsely chopped	125 mL

1. Combine cranberries, orange juice concentrate, sweet pepper, hot pepper, onion, garlic, vinegar, sugar, cumin and salt in a medium stainless steel or enamel saucepan. Bring to a boil over high heat, reduce heat and cook gently, uncovered, for 15 minutes or until mixture thickens and vegetables are soft.
2. Remove hot jars from canner. Stir cilantro into salsa; ladle salsa into jars to within ½ inch (1 cm) of rim (head space). Process for 10 minutes for half-pint (250 mL) jars and 15 minutes for pint (500 mL) jars as directed on page 98 (Longer Time Processing Procedure).

Makes 2 cups (500 mL).

Tip: Use fresh and frozen cranberries interchangeably. Any time sugar is used, remember that too much will detract from cranberries' distinctive flavour.

Papaya Mango Salsa

This is a true fruit salsa and it's processed to keep on the shelf. Its papaya, mango and pineapple flavours go particularly well with fish.

1	papaya, peeled, seeded and chopped	1
1	mango, peeled and chopped	1
1	jalapeño pepper, seeded and finely chopped	1
	Juice and grated rind of 1 lime	
¼ cup	unsweetened pineapple juice	50 mL
1 tbsp	finely chopped crystallized ginger	15 mL
1 tbsp	rice wine vinegar	15 mL
¼ tsp	pickling salt	1 mL
2 tbsp	chopped fresh mint	25 mL

1. Place papaya, mango, jalapeño pepper, lime juice and rind, pineapple juice, ginger, vinegar and salt in a medium saucepan. Bring to a boil over high heat, reduce heat and boil gently for 1 minute. Stir in mint, return to a boil and cook for 1 minute.
2. Remove hot jars from canner and ladle salsa into jars to within ½ inch (1 cm) of rim (head space). Process for 20 minutes for half-pint (250 mL) jars and for pint (500 mL) jars as directed on page 98 (Longer Time Processing Procedure).

Makes 2 cups (500 mL).

Salsa Bruschetta-Style

Bruschetta has its origins in ancient Rome. During the December and January holidays, Romans celebrated by eating flat buns soaked in olive oil still fresh from the fall harvest. The name bruschetta comes from the Italian *bruscare*, meaning "to roast over coals." This topping is a favourite of ours to have on hand for making a quick appetizer.

3 cups	chopped peeled Italian plum tomatoes (1¼ lb/625 g)	750 mL
2	large cloves garlic, minced	2
2	shallots, minced	2
1 cup	chopped fresh basil	250 mL
1 tbsp	red wine vinegar	15 mL
1 tsp	lemon juice	5 mL
½ tsp	pickling salt	2 mL
¼ tsp	coarsely ground black pepper	1 mL
2	green onions, minced	2
3 tbsp	tomato paste	45 mL

1. Combine tomatoes, garlic, shallots, basil, vinegar, lemon juice, salt and pepper in a large stainless steel or enamel saucepan. Bring to a boil over high heat, reduce heat and boil gently for 5 minutes, stirring frequently. Stir in green onion, tomato paste and return to a boil.
2. Remove hot jars from canner and ladle salsa into jars to within ½ inch (1 cm) of rim (head space). Process for 35 minutes for half-pint (250 mL) jars and 40 minutes for pint (500 mL) jars as directed on page 98 (Longer Time Processing Procedure).

Makes 3 cups (750 mL).

Bruschetta
These popular appetizers can be ready at a moment's notice with this salsa on hand.
Toast sliced Italian bread and rub with cut surface of a garlic clove. Brush lightly with olive oil and spoon on salsa. If desired, sprinkle with Parmesan cheese and place under a broiler for several minutes to warm.

Fresh Vegetable Salsas

Vegetable salsas are the everyday salsas of the salsa-eating world. Made and used fresh, or refrigerated for a short time, these salsas do not require as much vinegar as those that are canned and stored for later use.

Pico de Gallo Salsa

Pico de Gallo is a blend of hot chile peppers and other fresh vegetables. Variations of this salsa can be found on every Mexican-restaurant table to eat with crisp corn or flour tortillas. It is excellent as a dip, and also as a salsa on tacos and enchiladas and with grilled meats and poultry. We hope it will become as popular in your home as in ours.

4	plum tomatoes, chopped (about 2 cups/500 mL)	4
½ cup	chopped red onion	125 mL
½ cup	chopped cucumber	125 mL
1	large clove garlic, minced	1
1	small jalapeño pepper, seeded and minced	1
¼ cup	chopped fresh cilantro	50 mL
¼ cup	lime juice	50 mL

1. Combine tomatoes, onion, cucumber, garlic, jalapeño pepper, cilantro and lime juice. Stir well.
2. Cover and refrigerate for 30 minutes (longer is unnecessary as this salsa is best when freshly made).

Makes about 3 cups (750 mL).

Fresh Tomato and Black Olive Salsa

The olives convert this basic Mexican dish to something more Spanish or maybe Italian. It's good as an appetizer with crackers or squares of toasted Italian bread. It is best served at room temperature.

2 cups	diced plum tomatoes (about 4)	500 mL
½ cup	chopped pitted black olives (preferably Kalamata)	125 mL
⅓ cup	chopped red onion	75 mL
2 tbsp	red wine vinegar	25 mL
1	clove garlic, minced	1
1 tbsp	Dijon mustard	15 mL
	Freshly ground black pepper	

1. Combine tomatoes, olives, onion, vinegar, garlic, mustard and pepper to taste. Stir well.
2. Cover and let stand at room temperature for a few hours. Or place in tightly sealed containers and refrigerate up to 2 weeks.

Makes about 2½ cups (625 mL).

Tomato Freezer Salsa

Take advantage of all those wonderful field tomatoes available in late summer by making this freezer salsa. Enjoy some before you freeze it.

5	large field tomatoes, peeled	5
2 tbsp	olive oil	25 mL
6	green onions, finely chopped	6
1	jalapeño pepper, seeded and finely chopped	6
4	medium cloves garlic, finely chopped	4
¼ cup	chopped fresh cilantro	50 mL
2 tbsp	lime juice	25 mL
¼ tsp	each salt and freshly ground pepper	1 mL

1. Coarsely chop tomatoes and place in a sieve over a bowl to allow extra juice to drain.
2. Heat oil on medium heat in a nonstick skillet. Add onions, jalapeño pepper and garlic; cook for 7 minutes or until softened but not brown, stirring often. Let cool. Stir into drained tomatoes.
3. Stir in lime juice, salt and pepper. Spoon into tightly sealed plastic containers and freeze for up to 4 months.

Makes about 5 cups (1.25 L).

Fresh Fruit Salsas

Although vegetable salsas are by far the most common, trendy fresh fruit salsas are appearing more frequently. Their uses are many—and everyone eats large amounts as a side dish, so make lots. They are refreshing.

Fresh Spicy Tropical Fruit Salsa

Tropical flavours abound in this fresh fruit salsa. We found that its fresh and minty flavours beautifully complement grilled sausages, pork chops, pork tenderloin and chicken.

1	kiwifruit, peeled and diced	1
¼	mango, peeled and diced	¼
¼	papaya, peeled, seeded and chopped	¼
½ cup	quartered strawberries	125 mL
½ cup	diced cantaloupe	125 mL
½	jalapeño or other hot pepper, seeded and finely chopped	½
2 tbsp	finely chopped fresh mint	25 mL
1 tbsp	each granulated sugar and lime juice	15 mL

1. Gently stir together kiwifruit, mango, papaya, strawberries, cantaloupe and jalapeño. Add mint, sugar and lime juice; stir to blend. Refrigerate for about 10 minutes to allow flavours to develop.

Makes 3 cups (750 mL).

Fresh Sweet Pepper and Peach Salsa

Peaches are to summer as apples are to fall, and this salsa is certainly a celebration of the summer peach season. It is also marvellous made with nectarines. Serve it with pork tenderloin or barbecued pork chops. We also like it with chicken breasts and chicken burgers.

4 cups	chopped peeled peaches (4 medium)	1 L
½	small sweet red pepper, chopped	½
½	small sweet green pepper, chopped	½
¼ cup	finely chopped red onion	50 mL
¼ cup	chopped fresh cilantro	50 mL
1 tbsp	chopped jalapeño or other hot pepper	15 mL
1	clove garlic, crushed	1
1 tbsp	each lime juice and rice vinegar	15 mL
1 tsp	liquid honey	5 mL

1. Combine peaches, sweet peppers, onion, cilantro, jalapeño pepper and garlic in a medium bowl.
2. Stir in lime juice, vinegar and honey. Cover and refrigerate for 30 minutes for flavours to develop.

Makes about 3½ cups (875 mL).

Fresh Pineapple Jalapeño Salsa

Hot jalapeño pepper and sweet, cooling pineapple blend to produce a salsa to serve with grilled chicken or fish. We did, and everyone wanted the recipe.

3	plum tomatoes, diced	3
1 cup	diced pineapple	250 mL
1 cup	diced papaya	250 mL
¼ cup	chopped fresh cilantro	50 mL
2	green onions, chopped	2
2	small jalapeño peppers, seeded and finely chopped	1
2 tbsp	lime juice	25 mL
Pinch	each salt and freshly ground pepper	Pinch

1. Combine tomatoes, pineapple, papaya, cilantro, onions and jalapeño in a bowl.
2. Stir in lime juice, salt and pepper. Cover and refrigerate for 3 hours before serving. Stir again and transfer to a serving bowl.

Makes 2½ cups (625 mL).

Southwest Black Bean and Corn Salsa

Cilantro, cumin and jalapeño peppers deliver flavours reminiscent of the Southwest.

1 cup	cooked black beans	250 mL
1 cup	frozen or fresh corn niblets	250 mL
½ cup	finely chopped celery	125 mL
2	jalapeño peppers, seeded and finely chopped	2
2	cloves garlic, chopped	2
2 tbsp	each lime juice, balsamic vinegar	25 mL
1 tbsp	red wine vinegar	15 mL
¼ tsp	each cumin, coarsely ground black pepper and salt	1 mL
½ cup	each finely chopped sweet red pepper and green onion	125 mL
1 tbsp	olive oil	15 mL
½ cup	chopped fresh cilantro	125 mL

1. Place beans, corn, celery, jalapeño peppers, garlic, lime juice, balsamic vinegar, red wine vinegar, cumin, pepper and salt in a large saucepan. Bring to a boil, reduce heat and boil gently for 5 minutes.
2. Stir in red pepper, green onion, olive oil and cilantro.
3. Spoon salsa into clean jars or plastic containers to within ½ inch (1 cm) of rim. Cover with tight-fitting lids. Label jars and refrigerate for up to 1 week or freeze for longer storage.

Makes 2¾ cups (675 mL).

Tip: To Cook Beans: Place washed dried beans in a large saucepan and add 3 times the amount of cold water. Cover and bring to a boil; boil for 2 minutes. Remove from heat and let stand for 1 hour; drain. Cover beans with cold water and bring to a boil; reduce heat, cover and boil gently, covered, for 30 minutes or until beans are tender; drain.

Tapenade-Style Salsa

This salsa has a more paste-like consistency than most and with its olive and garlic flavours reminds us of the French tapenade. It is marvellous spread on sliced crusty Italian bread or on crackers.

¾ cup	chopped pitted Kalamata olives	175 mL
⅔ cup	chopped pitted green olives	150 mL
½	sweet red pepper, chopped	½
¼ cup	chopped sun-dried tomatoes in olive oil	50 mL
¼ cup	chopped Italian (flat-leaf) parsley	50 mL
¼ cup	olive oil	50 mL
2	cloves garlic, chopped	2
1 tbsp	each red wine vinegar and balsamic vinegar	15 mL
	Freshly ground black pepper	

1. Place olives, red pepper, dried tomatoes, parsley, oil, garlic, red wine vinegar and balsamic vinegar in a food processor and process with on/off motion until finely chopped. Add black pepper to taste.
2. Spoon salsa into clean jars or plastic containers to within ½ inch (1 cm) of rim. Cover with tight-fitting lids. Label jars and refrigerate for up to 1 week or freeze for longer storage.

Makes 2 cups (500 mL).

Chapter Nine

Choice
Chutneys

C HUTNEYS are rich-tasting condiments containing a wide variety of fruits and vegetables along with vinegar, sugar and spices to give them their characteristic sweet-sour flavour. Chutneys are smooth yet pulpy, mellow, very full flavoured *and* easily prepared.

Traditionally, apples and onions are the base chutney ingredients, with raisins and sometimes dates added. Tomatoes also make marvellous chutneys, either fresh tomatoes in season or canned. Chutneys range from mild to hot in spiciness. Chile peppers, peppercorns and mustard seeds provide the spicing "bang" and, along with such other spices as cinnamon, allspice and ginger, intensify the flavour. Garlic and shallots add extra zest.

Generally, chutneys need long, steady cooking in an open saucepan to reach the correct consistency. But do not overcook them. As soon as the mixture begins to thicken to the point where pools of liquid no longer collect on the surface, the chutney is ready to ladle into jars for processing. Try to resist opening the jars for several weeks. Chutneys truly do mellow and improve in flavour as they age.

Serving Suggestions

Chutneys are traditionally served as an accompaniment to hot meals like curries, rice dishes, stews and casseroles. But don't hesitate to pair them with roast chicken, lamb, beef, pork and game. Some greatly enhance fish. Offer a selection of chutneys with barbecued meats as well as with salads and cold meats. We often serve them with such cheeses as Stilton, Brie, Camembert and gorgonzola. And be sure to try them with a cottage cheese salad. The sweeter ones make delicious bread and cracker spreads. One day, a small amount of chutney remained in the bottom of a jar when a dip was needed. We stirred in an equal amount of plain low-fat yogurt and *voilà*—the chutney became a delicious dip for raw veggies!

List of Recipes

Quick Apple Cranberry Chutney

This colourful recipe is a breeze! Cooking time is particularly short because of the high pectin content of both fruits. It's wonderfully spicy with lamb, chicken, pork and game.

2 cups	chopped cranberries	500 mL
1 cup	finely chopped apple	250 mL
½ cup	each finely chopped red onion and sweet red pepper	125 mL
½ cup	cider vinegar	125 mL
2	cloves garlic, minced	2
1 tbsp	finely chopped gingerroot	15 mL
½ cup	packed brown sugar	125 mL
¼ tsp	each cumin and salt	1 mL
Pinch	each freshly ground pepper and hot pepper flakes	Pinch

1. Combine cranberries, apple, onion, red pepper, vinegar, garlic and gingerroot in a stainless steel or enamel saucepan. Bring to a boil over high heat, reduce heat and boil gently, covered, for 5 minutes or until cranberries pop.
2. Add sugar, cumin, salt, pepper and hot pepper flakes. Cook for 5 minutes or until thickened.
3. Remove hot jars from canner and ladle chutney into jars to within ½ inch (1 cm) of rim (head space). Process for 10 minutes for half-pint (250 mL) jars and 15 minutes for pint (500 mL) jars as directed on page 98 (Longer Time Processing Procedure).

Makes 2 cups (500 mL).

To Microwave:

1. Combine cranberries, apple, onion, red pepper, vinegar, garlic and gingerroot in a shallow 4-cup (1 L) microwavable container. Microwave, covered, at High (100%) for 5 minutes or until cranberries pop.
2. Stir in sugar, cumin, salt, pepper and hot pepper flakes. Microwave, covered, at Medium-High (70%) for 5 minutes or until thickened, stirring once. Proceed from step 3 above.

All-Year Dried Fruit Chutney

All chutneys do not need to be prepared in the growing season. This one, using fruits available at any time of the year, is a marvellous addition to pork, poultry, beef and game. We have also used it as an appetizer with soft Brie or Camembert cheese on crackers.

1	large banana, mashed	1
1	large apple, peeled, cored and chopped	1
1	large red onion, chopped	1
3	cloves garlic, minced	3
½ cup	each chopped dried apricots, prunes and dates	125 mL
⅓ cup	chopped mixed glacéed fruit	75 mL
2 tbsp	chopped crystallized ginger	25 mL
1 cup	cider vinegar	250 mL
½ cup	water	125 mL
½ tsp	each cayenne, ground allspice, ground cardamom and salt	2 mL
1 cup	lightly packed dark brown sugar	250 mL

1. Combine banana, apple, onion, garlic, apricots, prunes, dates, glacéed fruit, ginger, vinegar and water in a large stainless steel or enamel saucepan. Bring to a boil over high heat, reduce heat and boil gently, uncovered, for 10 minutes, stirring occasionally. Add cayenne, allspice, cardamom, salt and sugar; boil gently, stirring occasionally, for 10 minutes or until thickened.
2. Remove hot jars from canner and ladle chutney into jars to within ½ inch (1 cm) of rim (head space). Process for 10 minutes for half-pint (250 mL) jars and 15 minutes for pint (500 mL) jars as directed on page 98 (Longer Time Processing Procedure).

Makes 4 cups (1 L).

Indian Chutney

Along the lines of the commercial chutney known as Major Grey, this recipe has a somewhat zippier flavour than some of our others. We love it with curries and also as a light appetizer combined with low-fat sour cream or plain yogurt served with crackers or raw vegetables.

1 cup	chopped onion	250 mL
¾ cup	raisins	175 mL
¾ cup	cider vinegar	175 mL
1	medium orange, peeled and chopped	1
1	medium lemon, peeled and chopped	1
1	lime, peeled and chopped	1
¼ cup	each lightly packed brown sugar and molasses	50 mL
¼ cup	finely chopped gingerroot	50 mL
4	cloves garlic, crushed	4
1 tbsp	mustard seeds	15 mL
½ tsp	each hot pepper flakes and cinnamon	2 mL
¼ tsp	each ground cloves and allspice	1 mL
Pinch	cayenne pepper	Pinch

1. Combine onion, raisins, vinegar, orange, lemon, lime, brown sugar, molasses, gingerroot, garlic and mustard seeds in a large stainless steel or enamel saucepan. Bring to a boil over high heat, reduce heat and boil gently, uncovered, for 30 minutes or until fruit is tender and mixture is thickened, stirring occasionally. Add hot pepper flakes, cinnamon, cloves, allspice and cayenne; boil gently for 5 minutes.
2. Remove hot jars from canner and ladle chutney into jars to within ½ inch (1 cm) of rim (head space). Process for 10 minutes for half-pint (250 mL) jars and 15 minutes for pint (500 mL) jars as directed on page 98 (Longer Time Processing Procedure).

Makes 3 cups (750 mL).

Indian Chutney Cheese Spread: Blend ⅓ cup (75 mL) chutney with ⅔ cup (150 mL) light cream cheese. Use as a spread for crackers or to fill celery sticks and hollowed-out cherry tomatoes. Makes 1 cup (250 mL).

Spiced Kiwifruit Apple Chutney

We enjoy making this chutney when few other fresh fruits are available. Try it as an appetizer with cream cheese. Its delicate flavour pairs well with fish.

7	kiwifruit, peeled and chopped (about 3 cups/750 mL)	7
2	apples, peeled, cored and chopped	2
¾ cup	finely chopped onion	175 mL
¾ cup	granulated sugar	175 mL
¾ cup	cider vinegar	175 mL
¼ cup	brown sugar	50 mL
⅓ cup	golden raisins	75 mL
2	cloves garlic, crushed	2
1 tsp	minced peeled gingerroot	5 mL
½ tsp	each cinnamon and mustard seeds	2 mL
¼ tsp	each cayenne pepper, ground cloves, nutmeg and salt	1 mL

1. Combine kiwifruit, apples, onion, granulated sugar, vinegar, brown sugar, raisins, garlic and gingerroot in a large stainless steel or enamel saucepan. Bring to a boil over high heat, reduce heat and boil gently, uncovered, for 25 minutes or until thickened and fruit is tender, stirring occasionally. Add cinnamon, mustard seeds, cayenne, cloves, nutmeg and salt; boil gently for a few minutes longer.
2. Remove hot jars from canner and ladle chutney into jars to within ½ inch (1 cm) of rim (head space). Process for 10 minutes for half-pint (250 mL) jars and 15 minutes for pint (500 mL) jars as directed on page 98 (Longer Time Processing Procedure).

Makes 4 cups (1 L).

Chicken Salad with Chutney Cream Dressing
Dressing: Combine in a blender ½ cup (125 mL) plain yogurt or light sour cream, ¼ cup (50 mL) Spiced Kiwifruit Apple Chutney, 2 tbsp (25 mL) light mayonnaise, 1 tbsp (15 mL) Dijon mustard and 1 tsp (5 mL) dry sherry; process until smooth. Makes about 1 cup (250 mL).
Salad: Combine 1 cup (250 mL) diced cold cooked chicken and ½ cup (125 mL) each chopped celery and apple in a bowl. Toss with enough dressing to moisten.

Mango Chutney

Mango Chutney is the one we think of as the "original" and most traditional of all chutneys. It goes well with curries, chicken, pork, lamb and game. See also the Mango Chutney Vinaigrette on page 221.

3	medium apples, peeled, cored and chopped	3
2	large mangoes, peeled and chopped	2
½	medium sweet red pepper, chopped	½
1½ cups	granulated sugar	375 mL
1 cup	finely chopped onion	250 mL
½ cup	golden raisins	125 mL
½ cup	white vinegar	125 mL
¼ cup	finely chopped peeled gingerroot	50 mL
1 tbsp	lemon juice	15 mL
2 tsp	curry powder	10 mL
½ tsp	each ground nutmeg, cinnamon and salt	2 mL

1. Combine apples, mangoes, red pepper, sugar, onion, raisins, vinegar and ginger-root in a large stainless steel or enamel saucepan. Bring to a boil over high heat, reduce heat and boil gently, uncovered, for 20 minutes or until fruit is tender and mixture is thickened, stirring occasionally. Add lemon juice, curry powder, nutmeg, cinnamon and salt; boil gently for 5 minutes.

2. Remove hot jars from canner and ladle chutney into jars to within ½ inch (1 cm) of rim (head space). Process for 10 minutes for half-pint (250 mL) jars and 15 minutes for pint (500 mL) jars as directed on page 98 (Longer Time Processing Procedure).

Makes 5 cups (1.25 L).

Chutney Butter
Serve with grilled or barbecued chicken parts.
Combine 3 tbsp (45 mL) Mango Chutney and 1 tbsp (15 mL) softened butter or margarine. Stir in 2 tsp (10 mL) chopped fresh cilantro and a pinch of cayenne pepper. Makes ¼ cup (50 mL).

Mixed Fruit Chutney

This zesty chutney was shared with us by Sheila Whyte of Thyme and Again Creative Catering and Take Home Foods in Ottawa and her chef, Robert Jutres. Robert created this recipe one fall when apples, pears and plums were abundant. It has become one of their best-sellers. Robert suggests serving it with grilled chicken and curries. He says, "Don't be tempted to taste it for at least 2 weeks. Allow the flavours to mellow."

6 cups	diced cored peeled tart apples (about 2 lb/1 kg)	1.5 L
3 cups	coarsely chopped peeled tomatoes (about 1 lb/500 g)	750 mL
2 cups	diced cored peeled firm pears (about 1 lb/500 g)	500 mL
1 cup	diced prune plums (about ½ lb/250 g)	250 mL
1 cup	raisins or currants	250 mL
5 cups	lightly packed dark brown sugar	1.25 L
2½ cups	cider or malt vinegar	625 mL
½ tsp	each ground ginger, mace*, ground cloves, cayenne pepper, coarsely ground black pepper and salt	2 mL

1. Combine apples, tomatoes, pears, plums and raisins in a large stainless steel or enamel saucepan. Stir in brown sugar, vinegar, ginger, mace, cloves, cayenne pepper, black pepper and salt.
2. Bring to a boil over high heat, stirring constantly. Reduce heat and boil gently for 1 hour or until chutney is very thick and golden brown, stirring frequently.
3. Remove hot jars from canner and ladle chutney into jars to within ½ inch (1 cm) of rim (head space). Process for 15 minutes for half-pint (250 mL) jars and pint (500 mL) jars as directed on page 98 (Long Time Processing Procedure).

Makes 4½ cups (1.125 L).

Tip: Tomatoes peel easily after being dipped into boiling water for 30 seconds.

*Mace is the red covering of the nutmeg seed, dried and usually sold in a ground form. It has a fuller, spicier flavour than the nutmeg seed, but you may substitute ground nutmeg in recipes calling for mace.

Spicy Rhubarb Chutney

Unique among chutneys, this one offers the sour-sweet springtime taste of rhubarb. Its tartness goes particularly well with roast pork or ham, but also can be enjoyed with poultry.

4 cups	chopped rhubarb	1 L
1 cup	granulated sugar	250 mL
⅓ cup	white vinegar	75 mL
2	apples, peeled, cored and chopped	2
½ cup	raisins	125 mL
¼ cup	finely chopped onion	50 mL
1 tbsp	minced ground gingerroot	15 mL
1 tsp	each ground cinnamon and salt	5 mL
¼ tsp	ground cloves	1 mL

1. Combine rhubarb, sugar, vinegar, apples, raisins, onion and gingerroot in a medium stainless steel or enamel saucepan. Cook, covered, on medium heat for 10 minutes or until thickened and fruit is soft, stirring occasionally.
2. Add cinnamon, salt and cloves; cook for a few minutes longer, stirring frequently.
3. Remove hot jars from canner and ladle chutney into jars to within ½ inch (1 cm) of rim (head space). Process for 10 minutes for half-pint (250 mL) jars and 15 minutes for pint (500 mL) jars as directed on page 98 (Longer Time Processing Procedure).

Makes 4 cups (1 L).

For crisp, garlicky pickles, see
Sweet Garlic Dills on page 102.

Uncle Freddie's Tomato Chutney

Margaret's neighbour gave us this wonderful chutney recipe. He recommends serving it with beef, egg dishes or macaroni and cheese. It's a large year-round recipe, which may be halved with equally great results.

2	cans (28 oz/796 mL) chopped tomatoes	2
3	apples, peeled, cored and chopped	3
2	large onions, chopped	2
1 cup	cider vinegar	250 mL
¾ cup	golden raisins	175 mL
2 cups	lightly packed brown sugar	500 mL
1 tbsp	lemon juice	15 mL
2 tsp	salt	10 mL
1–2 tsp	curry powder	5–10 mL
½ tsp	each ground cinnamon, cloves and ginger	2 mL

1. Combine tomatoes, apples, onions, vinegar and raisins in a large stainless steel or enamel saucepan. Bring to a boil over high heat, reduce heat and boil gently, uncovered, for 30 minutes or until thickened, stirring occasionally.

2. Add sugar, lemon juice, salt, curry powder, cinnamon, cloves, and ginger. Return to a boil, reduce heat and boil gently for 15 minutes or until desired thickness, stirring frequently.

3. Remove hot jars from canner and ladle chutney into jars to within ½ inch (1 cm) of rim (head space). Process for 10 minutes for half-pint (250 mL) jars and 15 minutes for pint (500 mL) jars as directed on page 98 (Longer Time Processing Procedure).

Makes 8 cups (2 L).

Tip: If you want to use fresh tomatoes, replace each can with 5 large peeled and chopped tomatoes.

Fruit can be magically transformed
with the addition of wine or liqueur.
(See page 196.)

Chapter Ten

Savoury Sauces

SAVOURY SAUCES are piquant and full flavoured. They include mustards, ketchups and chili sauces.

Mustards are one of the all-time great condiments. The name comes from the Latin *mustum ardens*, meaning "burning wine," a concoction the Romans made by mixing unfermented grape juice (must) with mustard seeds. Down through the centuries, mustard has been cultivated and used both for medicinal as well as culinary purposes. It has become one of the world's most popular food flavourings.

Ketchups seem to be unique to North America. We have never found a ketchup or chili sauce recipe in a cookbook from any other country. An interesting observation, since a seventeenth-century Chinese condiment for fish called *ke-tsiap* is thought to be the origin of today's popular ketchup. English seamen took the sauce home, and by the late 1700s, it had travelled to New England, where tomatoes were added to the blend. Today we are able to find ketchups made from many interesting foods. Try our Microwave Mango Ketchup (page 165) and Piquant Tomato Sauce (page 174) and you may never go back to basic red.

Chili sauces are ketchup-like spicy sauces but with a coarser consistency, made with tomatoes, onions, green peppers, chile peppers or chili powder, vinegar, sugar and spices. We hope our Grandma's Chili Sauce (page 166) will bring back fond memories. Another favourite is Winter Mango Chili Sauce (page 167). And no cookbook would be complete without a Fruit Chili Sauce (page 168).

Asian cuisine uses many appealing sauces to accompany a great variety of dishes. Enjoy our easily prepared Asian Plum Sauce (page 171) and Chili Thai Sauce (page 172).

Serving Suggestions

Savoury sauces have many uses. After using our mustards on hamburgers, hotdogs and cold cuts, try them in a vinaigrette. Meat pies such as tourtière and meat loaf would be incomplete without chili sauce.

List of Recipes

Mustards

Mustard has been used since Roman times and commercial makers offer us many varieties. The mustard seed is usually ground before being mixed with a liquid to mellow its natural bitterness. Then the mustard is cooked at length to decrease the pungency and aged to develop the flavours. Commercial makers jealously guard their secret recipes, but you can make a delectable alternative that will keep for months in your refrigerator.

Basic Coarse Mustard

Using mustard seeds instead of mustard powder gives the interesting coarse texture to this condiment. This recipe can be used in its basic form or in its several variations.

⅓ cup	mustard seeds	75 mL
⅓ cup	cider vinegar	75 mL
1	clove garlic, halved	1
3 tbsp	water	45 mL
3 tbsp	liquid honey	45 mL
¼ tsp	salt	1 mL
Pinch	ground cinnamon	Pinch

1. Combine mustard seeds, vinegar and garlic in a small bowl. Cover and refrigerate for 36 hours.
2. Discard garlic. Process mixture in a food processor with water until coarse consistency. Stir in honey, salt and cinnamon.
3. Divide mixture into 3 equal parts and proceed as below.

Horseradish Mustard: To ⅓, add 1 tsp (5 mL) horseradish.
Peppercorn Mustard: To ⅓, add 1 tsp (5 mL) green peppercorns, crushed.
Herb Mustard: To ⅓, add ¼ tsp (1 mL) or more as desired of any dried herb. (Try tarragon, dill, thyme, basil or oregano.)

Refrigerate in tightly sealed containers.

Makes ¼ cup (50 mL) of each variation.

Dijonnaise Mustard

Try this on sandwiches with cold cuts, ham or roast beef.

Add ⅓ cup (75 mL) mayonnaise to 2 tbsp (25 mL) Basic Coarse Mustard. Makes about ½ cup (125 mL).

Creamy Mustard Sauce

Use on cooked vegetables, to top baked potatoes and to dress a cabbage or tossed green salad.

To ¼ cup (50 mL) Horseradish Mustard, add ½ cup (125 mL) light sour cream, 2 tbsp (25 mL) lemon juice and 2 tbsp (25 mL) chopped fresh parsley. Makes about ¾ cup (175 mL).

Wine Mustard

Legend has it that a vinegar and mustard maker in Dijon first added wine to his mustard in the 1700s, and so began the resurgence of mustards in Europe.

½ cup	liquid honey	125 mL
⅓ cup	white wine	75 mL
¼ cup	dry mustard	50 mL
1	egg	1
1 tbsp	vegetable oil	15 mL
1 tsp	all-purpose flour	5 mL

1. Combine honey, wine and dry mustard in a small saucepan. Whisk in egg, oil and flour; cook over medium heat for 2 to 3 minutes or until bubbly, stirring constantly. Cook for 1 minute longer, stirring constantly. Remove from heat.
2. Cool to room temperature. Cover and refrigerate in a tightly sealed container for up to 2 weeks.

Makes 1 cup (250 mL).

Mustard Sauces

There are many uses for a great mustard sauce!

Roasted Red Pepper Mustard Sauce

An appetizer mustard for crackers with or without cheese. This sauce may also be reheated and served over cooked vegetables, pork or ham.

½ cup	finely chopped roasted sweet red peppers	125 mL
	OR bottled sweet red peppers	
	OR canned pimientos	
¼ cup	granulated sugar	50 mL
¼ cup	dry mustard	50 mL
¼ cup	red wine vinegar	50 mL
1 tbsp	butter or margarine	15 mL
¼ tsp	salt	1 mL
Pinch	freshly ground black pepper	Pinch
2 tbsp	water	25 mL
1 tsp	all-purpose flour	15 mL

1. Combine red peppers, sugar, mustard, vinegar, butter, salt and pepper in a small saucepan. Cook over medium heat for 5 minutes or until hot; stir constantly.
2. Stir together water and flour to make a smooth paste. Whisk into hot mixture. Cook over low heat until smooth and thickened, stirring frequently.
3. Cool sauce. Refrigerate in a tightly sealed container for up to 1 month.

Makes 1 cup (250 mL).

To Roast Red Peppers
Roast sweet peppers in a 400°F (200°C) oven for 20 minutes or until blackened, turning several times. Place in a small paper bag until cool. Remove and discard skin and seeds.

Honey Mustard Sauce

Try the Mustard Appetizers (page 215) made with this sauce or use it in Honey Mustard Dressing (page 222) for a salad dressing or marinade.

½ cup	liquid honey	125 mL
⅓ cup	cider vinegar	75 mL
¼ cup	dry mustard	50 mL
1 tbsp	vegetable oil	15 mL
1 tbsp	molasses	15 mL
¼ tsp	each ground allspice, ground cloves and salt	1 mL
Pinch	freshly ground pepper	Pinch
1 tbsp	all-purpose flour	15 mL
2 tbsp	water	25 mL

1. Combine honey, vinegar, mustard, oil, molasses, allspice, cloves, salt and pepper in a small saucepan. Cook over medium heat for 5 minutes or until hot; stir constantly.
2. Stir together flour and water to make a smooth paste. Whisk into hot mixture. Cook over low heat until smooth and thickened, stirring frequently.
3. Cool sauce. Refrigerate in a tightly sealed container for up to 1 month.

Makes about ⅔ cup (150 mL).

———————

Horseradish Honey Mustard
Delicious as a sauce for roast beef.
Combine ¼ cup (50 mL) Honey Mustard Sauce with 1 tbsp (15 mL) horseradish.

———————

Marmalade Honey Mustard Glaze
A glaze to dress up a ham before it goes to the table.
Combine equal amounts of Honey Mustard Sauce and one of the marmalades in chapter 3. Brush glaze on ham 15 to 30 minutes before the end of the roasting time.

Tip: See page 221 for salad dressing ideas.

Mustard Raisin Sauce

We especially like this sauce served with ham slices and grilled chicken.

½ cup	**Dijon mustard or Wine Mustard** (page 161)	125 mL
¼ cup	raisins	50 mL
¼ cup	maple syrup	50 mL
¼ cup	corn syrup	50 mL
¼ cup	orange juice	50 mL
2 tbsp	soy sauce	25 mL
1 tsp	cornstarch	5 mL
1 tbsp	water	15 mL
1 tsp	lemon juice	5 mL

1. Combine mustard, raisins, maple syrup, corn syrup, orange juice and soy sauce in a small saucepan. Bring to a boil over high heat, reduce heat and simmer for 1 minute, stirring constantly.
2. Mix together cornstarch and water. Whisk into hot mustard mixture. Cook over low heat until smooth and thickened, for 1 minute, stirring frequently. Stir in lemon juice.
3. Cool sauce. Refrigerate in a tightly sealed container for up to 2 weeks.

Makes 1⅔ cups (400 mL).

Microwave Mango Ketchup

Ketchup hasn't always been made from tomatoes. This popular condiment origi-
nated in seventeenth-century China, when it was made of spicy pickled fish. British
seamen took it home and later added tomatoes, making the blend we know today.
Try this interesting version on chicken or beef burgers or blackened fish.

2	mangoes, peeled and finely chopped	2
¼ cup	granulated sugar	50 mL
¼ cup	dry white wine	50 mL
¼ cup	cider vinegar	50 mL
1 tsp	ground ginger	5 mL
½ tsp	salt	2 mL
¼ tsp	each ground allspice and cloves	1 mL

1. Combine mangoes, sugar, wine, vinegar, ginger, salt, allspice and cloves in a
   small microwavable container. Microwave, uncovered, on High (100%) for 5
   minutes. Stir. Microwave on Low (30%) for 3 to 5 minutes or until mixture is
   very thick, stirring several times.
3. Remove hot jars from canner and ladle ketchup into jars to within ½ inch
   (1 cm) of rim. Process for 15 minutes for half-pint (250 mL) jars and 20 minutes for pint (500 mL) jars as directed on page 98 (Longer Time Processing
   Procedure).

Makes 2 cups (500 mL).

Grandma's Chili Sauce

There are probably as many recipes for chili sauce as there are grandmothers. We especially like this recipe, shared with us by a Hamilton home economist. It was her grandmother's specialty.

4 cups	diced peeled tomatoes	1 L
5	stalks celery, finely diced	5
2	apples, peeled, cored and diced	2
1	small sweet green pepper, finely diced	1
1	small hot red pepper, seeded and finely chopped	1
1	small onion, finely chopped	1
½	sweet red pepper, finely diced	½
1 cup	white vinegar	250 mL
⅓ cup	granulated sugar	75 mL
½ tsp	pickling salt	2 mL
3	cinnamon sticks, each 3 inches (8 cm) long	3
1	1-inch (2.5 cm) piece dried whole ginger	1
1 tsp	whole allspice berries	5 mL

1. Place tomatoes, celery, apples, green pepper, hot pepper, onion, red pepper, vinegar, sugar and salt in a large stainless steel or enamel saucepan.
2. Tie cinnamon, ginger and allspice in a small square of cheesecloth. Add to vegetables. Bring to a boil over high heat, reduce heat and boil gently, uncovered, for about 1½ hours or until mixture is thick, stirring occasionally. Discard spice bag.
3. Remove hot jars from canner and ladle sauce into jars to within ½ inch (1 cm) of rim (head space). Process for 15 minutes for half-pint (250 mL) and pint (500 mL) jars as directed on page 98 (Longer Time Processing Procedure).

Makes 4 cups (1 L).

Winter Mango Chili Sauce

Fall isn't the only time for making chili sauce. Mangoes blend beautifully with canned tomatoes to make this super-easy condiment.

1	can (28 oz/796 mL) plum tomatoes	1
1	large onion, chopped	1
1	sweet green pepper, chopped	1
1	mango, peeled and chopped	1
1 cup	white vinegar	250 mL
½ tsp	each pickling salt and ground cinnamon	2 mL
¼ tsp	ground cloves	1 mL
Pinch	cayenne pepper	Pinch
⅓ cup	granulated sugar	75 mL

1. Combine tomatoes, onion, green pepper, mango, vinegar, salt, cinnamon, cloves and cayenne in a large stainless steel or enamel saucepan. Bring to a boil over high heat, reduce heat and boil gently, uncovered, for about 1½ hours or until thick, stirring occasionally.
2. Stir sugar into sauce and return to a boil.
3. Remove hot jars from canner and ladle sauce into jars to within ½ inch (1 cm) of rim (head space). Process for 15 minutes for half-pint (250 mL) and pint (500 mL) jars as directed on page 98 (Longer Time Processing Procedure).

Makes 4 cups (1 L).

Fruit Chili Sauce

This traditional relish originates from the fruit-growing regions of Canada where peaches and pears are abundant in the fall.

6 cups	chopped peeled tomatoes	1.5 L
2 cups	finely chopped onions	500 mL
2 cups	each chopped peeled peaches, pears and apples	500 mL
1 cup	finely chopped celery	250 mL
2	sweet green peppers, finely chopped	2
1	sweet or hot red pepper, seeded and finely chopped	1
1½ cups	white or cider vinegar	375 mL
1 tbsp	pickling spice	15 mL
2 tsp	pickling salt	10 mL
2 cups	granulated sugar	500 mL

1. Combine tomatoes, onions, peaches, pears, apples, celery, green and red peppers and vinegar in a large stainless steel or enamel saucepan. Place pickling spice in a tea ball or tie in a piece of cheesecloth. Add spice and salt to saucepan. Bring to a boil over high heat, reduce heat and boil gently, uncovered, for 1 hour or until thick, stirring occasionally.
2. Stir in sugar. Return to a boil and boil gently for 30 minutes, stirring occasionally. Remove spice bag.
3. Remove hot jars from canner and ladle sauce into jars to within ½ inch (1 cm) of rim (head space). Process for 15 minutes for half-pint (250 mL) and pint (500 mL) jars as directed on page 98 (Longer Time Processing Procedure).

Makes 8 cups (2 L).

Variation:

For a spicier version, double the pickling spice and add 1 tsp (5 mL) whole cloves and 1 broken cinnamon stick to the spice bag.

Cranberry Sauce with Spirit

There will be no going back to traditional cranberry sauce once you have tried cranberries cooked with port wine.

1 cup	granulated sugar	250 mL
¼ cup	water	50 mL
1 tbsp	red wine vinegar	15 mL
2½ cups	fresh or frozen cranberries	625 mL
½ cup	port	125 mL
2	cinnamon sticks, each 3 inches (8 cm) long	2

1. Combine sugar, water and vinegar in a large stainless steel or enamel saucepan. Bring to a boil over high heat, stirring to dissolve sugar. Add cranberries; return to a boil, reduce heat and boil gently, uncovered, for 5 minutes, stirring frequently. Stir in port.
2. Remove sterilized jars from canner and place a cinnamon stick in each jar. Ladle sauce into jars to within ½ inch (1 cm) of rim (head space). Process for 15 minutes for half-pint (250 mL) and pint (500 mL) jars as directed on page 98 (Longer Time Processing Procedure).

Makes 2 cups (500 mL).

Herbed Raspberry and Red Currant Sauce

This great fruit combo provides lots of flavour interest to grilled or barbecued chicken breasts and pork chops. Stirred into plain yogurt, it makes an excellent fruit salad topping.

2 cups	fresh or frozen unsweetened raspberries	500 mL
2 cups	fresh or frozen red currants	500 mL
½ cup	water	125 mL
1½ cups	granulated sugar	375 mL
½ tsp	each dried tarragon and thyme	2 mL

1. Combine raspberries, currants and water in a medium stainless steel or enamel saucepan. Bring to a boil over high heat, reduce heat and boil gently, covered, for 20 minutes. Strain mixture through a fine sieve or cloth; discard pulp.
2. Return sauce to pan, return to a boil and slowly add sugar, stirring constantly until sugar is dissolved. Stir in tarragon and thyme; boil gently for 5 minutes.
3. Remove hot jars from canner and ladle sauce into jars to within ½ inch (1 cm) of rim (head space). Process for 10 minutes for half-pint (250 mL) jars and pint (500 mL) jars as directed on page 98 (Longer Time Processing Procedure).

Makes 3 cups (750 mL).

Asian Plum Sauce

Sauces of this type are not only difficult to find but often expensive. The Asian flavours complement roast pork, meatballs and of course such Oriental foods as spring and egg rolls. Margaret's daughter Janice, who gave us this recipe, uses it as a dipping sauce for cheese bites and sausage rolls.

9	purple plums, washed and pitted (about 1½ lb/750 g)	9
1½ cups	firmly packed brown sugar	375 mL
1 cup	cider vinegar	250 mL
1½ tsp	salt	7 mL
1½ cups	finely chopped onion	375 mL
3	cloves garlic, crushed	3
¼ cup	raisins	50 mL
2 tsp	soy sauce	10 mL
¼ tsp	chili powder	1 mL
Pinch	each ground cloves, cinnamon, ginger and allspice	Pinch

1. Finely chop plums in a food processor or by hand. You should have about 1¾ cups (425 mL).
2. Combine plums, sugar, vinegar and salt in a large stainless steel or enamel saucepan. Bring to a boil over high heat and boil gently, uncovered, for 3 minutes, stirring occasionally.
3. Add onion, garlic, raisins, soy sauce, chili powder, cloves, cinnamon, ginger and allspice to saucepan. Return to a boil, reduce heat and boil gently, uncovered, for 45 minutes or until mixture is thickened, stirring occasionally.
4. Remove hot jars from canner and ladle sauce into jars to within ½ inch (1 cm) of rim (head space). Process for 10 minutes for half-pint (250 mL) jars and for pint (500 mL) jars as directed on page 98 (Longer Time Processing Procedure).

Makes 3½ cups (875 mL).

Chili Thai Sauce

Thai flavours of fish sauce, lime and garlic combine to make this amazing sauce to serve with fish or chicken. It gives an irresistible flavour to Thai Chicken Wings (page 215).

1	small tomato, chopped	1
½	small sweet red pepper, chopped	½
½ cup	chopped onion	125 mL
1	clove garlic, minced	1
3 tbsp	minced gingerroot	45 mL
½ cup	chicken stock	125 mL
⅓ cup	fish sauce	75 mL
3 tbsp	lime juice	45 mL
2 tbsp	each brown sugar and rice vinegar	25 mL
2 tsp	hot pepper flakes	25 mL
¼ cup	chopped fresh cilantro	50 mL

1. Combine tomato, pepper, onion, garlic, gingerroot, chicken stock, fish sauce, lime juice, sugar, vinegar and hot pepper flakes in a medium saucepan. Bring to a boil over high heat, reduce heat and boil gently, uncovered, for 10 minutes. Remove from heat and stir in cilantro.
2. Remove hot jars from canner and ladle sauce into jars to within ½ inch (1 cm) of rim. Process for 15 minutes for half-pint (250 mL) jars and for pint (500 mL) jars as directed on page 98 (Longer Time Processing Procedure).

Makes 2 cups (500 mL).

Thai Baked Fish
Spread several spoonfuls of Chili Thai Sauce over fish fillets such as orange roughy, halibut or salmon. Bake in a 400°F (200°C) oven for 12 minutes or until fish is opaque and flakes easily with a fork. Serve with additional sauce.

Indonesian Satay Sauce

Fabulous as a sauce for pork kabobs or satays, this sauce also makes a delightful dip for raw vegetables.

2 tsp	vegetable oil	10 mL
1 cup	finely chopped onion	250 mL
4	cloves garlic, minced	4
⅔ cup	cider vinegar	150 mL
½ cup	each molasses and soy sauce	125 mL
2 tsp	hot pepper flakes	10 mL
2 tsp	minced peeled gingerroot	10 mL
½ cup	peanut butter	125 mL

1. Heat oil in a nonstick saucepan over medium-high heat and sauté onion and garlic for 3 minutes or until tender, stirring frequently.
2. Add vinegar, molasses, soy sauce, hot pepper flakes and gingerroot. Bring to a boil, reduce heat and boil gently for 5 minutes. Blend in peanut butter; boil gently for 1 minute.
3. Remove hot jars from canner and ladle sauce into jars to within ½ inch (1 cm) of rim. Process for 20 minutes for half-pint (250 mL) jars and for pint (500 mL) jars as directed on page 98 (Longer Time Processing Procedure).

Makes 2 cups (500 mL).

Chicken Barbecued with Indonesian Satay Sauce
Indonesian Satay Sauce imparts a rich glaze to chicken. For faster barbecuing, microwave chicken on Medium (50%) until partially cooked before placing on hot grill.
Marinate chicken pieces in Indonesian Satay Sauce, covered, for 1 hour at room temperature or up to 6 hours in refrigerator. Remove chicken from marinade, reserving marinade, and place on hot barbecue grill. Cook on low heat for 45 minutes, turning several times and basting with reserved marinade. Alternatively, arrange chicken in a single layer on a foil-lined baking dish. Bake in a 375°F (190°C) oven for 30 to 35 minutes. Discard any leftover marinade.

Piquant Tomato Sauce

A dash of this tangy sauce adds zip to other sauces. Or use it to top meat loaves and ground beef patties. We like to keep a bottle in the refrigerator ready to put on the table to add zing to egg dishes, soups and sandwich fillings.

3 cups	coarsely chopped peeled tomatoes	750 mL
3	cloves garlic, minced	3
3	stalks celery, finely chopped	3
3	jalapeño peppers OR 1 long yellow hot wax pepper, seeded and finely chopped	3
1	medium onion, finely chopped	1
1 cup	water	250 mL
½ cup	cider vinegar	125 mL
¼ cup	each lightly packed brown sugar and tomato paste	50 mL
1 tsp	each chili powder and cumin	5 mL
½ tsp	each dry mustard and pickling salt	2 mL

1. Combine tomatoes, garlic, celery, jalapeño peppers, onion and water in a medium stainless steel or enamel saucepan. Bring to a boil, reduce heat, cover and boil gently for 20 minutes or until mixture is thickened and vegetables are soft, stirring occasionally. Cool slightly. Purée in a blender or food processor until smooth.
2. Return tomato mixture to saucepan. Stir in vinegar, sugar, tomato paste, chili powder, cumin, mustard and salt; boil gently for 5 minutes to blend flavours, stirring continuously. (The sauce has a tendency to "jump" from the pot.)
3. Remove hot jars from canner and ladle sauce into jars to within ½ inch (1 cm) of rim (head space). Process for 15 minutes for half-pint (250 mL) jars and 20 minutes for pint (500 mL) jars as directed on page 98 (Longer Time Processing Procedure).

Makes 2 pint (500 mL) jars.

Easy Seafood Sauce
Blend 1 cup Piquant Tomato Sauce, 1 tbsp (15 mL) horseradish and 1 tsp (5 mL) lemon juice. Makes 1 cup (250 mL).

Your Basic Multi-Use Tomato Sauce

This fabulous tomato-rich sauce is made with three kinds of tomatoes, sun-dried, plum and regular. We like to add it to soups and beef stews for a nice flavour boost.

10	plum tomatoes, peeled and chopped	10
10	large tomatoes, peeled and chopped	10
4	large cloves garlic, minced	4
2	large stalks celery, chopped	2
2	medium carrots, chopped	2
1	large onion, chopped	1
1	large zucchini, chopped	1
1	large sweet green pepper, chopped	1
½ cup	sun-dried tomatoes	125 mL
⅔ cup	dry red wine	150 mL
½ cup	red wine vinegar	125 mL
2	bay leaves	2
1 tbsp	pickling salt	15 mL
2 tsp	each dried oregano and basil	10 mL
1 tsp	granulated sugar	5 mL
¼ tsp	each ground cinnamon and pepper	2 mL
¼ cup	chopped fresh parsley	50 mL

1. Combine tomatoes, garlic, celery, carrots, onion, zucchini and green pepper in a large stainless steel or enamel saucepan. Add 1 cup (250 mL) water. Bring to a boil over high heat, reduce heat and boil gently, covered, for 25 minutes or until mixture begins to thicken, stirring occasionally.

2. Soak sun-dried tomatoes in boiling water until softened. Drain and dice. Add to sauce with wine, vinegar, bay leaves, salt, oregano, basil, sugar, cinnamon and pepper. Continue to boil gently until desired consistency, stirring frequently. Discard bay leaves and stir in parsley.

3. Remove hot jars from canner and ladle sauce into jars to within ½ inch (1 cm) of rim (head space). Process for 35 minutes for pint (500 mL) jars and 40 minutes for quart (1 L) jars as directed on page 98 (Longer Time Processing Procedure).

Makes 12 cups (3 L).

Presto Pesto

Make this quickly prepared sauce when lots of inexpensive basil is in the market or growing in your garden. Then freeze for future use. The uses for this sauce are many: add some to hot pasta; to oil and vinegar for a salad dressing; to mayonnaise for potato or cabbage salad; and to cream sauces for cooked vegetables when a hint of basil will be appreciated. Chapter 13 has more ideas for using pesto in appetizers.

2 cups	packed fresh basil leaves	500 mL
½ cup	packed fresh parsley	125 mL
½ cup	olive oil or Garlic Oil (page 190)	125 mL
2	garlic cloves, peeled	2
¼ cup	toasted pine nuts	50 mL
1 cup	grated Parmesan cheese	250 mL

1. Place basil and parsley in a food processor; process with on/off motion until finely chopped. Slowly add oil, garlic and pine nuts; process until blended. Transfer to a bowl and stir in cheese.
2. Spoon into three ½-cup (125 mL) containers, close tightly and refrigerate for up to 1 week or freeze for longer storage. One container will serve 2 to 3 people as a pasta sauce.

Makes about 1½ cups (375 mL) concentrated pesto.

Tip: Use Presto Pesto to make Pesto Pita Pizzas (page 214) and Pesto Torta Appetizer (page 216).

All Those Extras

Introduction to

All Those Extras

B Y "EXTRAS" we mean those foods that add that extra interest and enjoyment to our taste experiences. This section has three chapters. The first puts lids on specialty vinegars and flavoured oils. The second puts lids on desserts in a jar. The third takes lids off to do some nifty things with the contents of some of those jars now sitting on your pantry shelf.

Specialty vinegars are incredibly easy to make because their high acid level eliminates any need to process them in a hot-water canner. They add great flavours to a wide variety of dishes. Our vinegar recipes include herb ones, such as with chive blossoms or rosemary (page 185); fruit ones, such as strawberry, raspberry and Citron (page 183); garlic ones, such as Red Wine Oregano Garlic Vinegar (page 187) and a marvellous Provence-style one (page 188). Costing a fraction of what you would pay in the stores, they are indeed great little indulgences.

Flavoured oils get their flavours from added ingredients, often garlic and hot peppers. There is growing interest in the many possibilities offered by flavoured oils. Just a small amount in cooking or in marinades gives amazing results. When making your own flavoured oil, it is important to follow the recipe instructions carefully. There is a risk of botulism from improperly processed oils.

Even a perfect meal is enhanced with dessert, the "Finishing Touch." Our desserts come in a jar. Some are entrancing sauces to spoon over ice cream or frozen yogurt. Hazelnut Blueberry Mango Sauce (page 199) is an excellent example, as is Cherry Compote (page 198). Others are fruits in liqueurs, like Brandied Dried Fruit Preserves (page 200). The chapter also includes some fruit liqueur recipes to capture the essence of a fruit for the final finishing touch.

There are endless ways to use preserved foods. We show you just a few in the final chapter "Let's Open the Lids and Use Them." Our suggestions include Spiced Plum Butter in a bran muffin (page 213), cheese spreads (page 216), and salad dressings (pages 221–222). We know you will think of many others.

Chapter Eleven

Specialty Vinegars and Flavoured Oils

Vinegars

Bottles of pastel vinegars containing delicate sprigs of herbs, spices or garlic make one of the most beautiful pictures in all of food land. Fresh herbs are steeped in vinegar, then stored in clear attractive glass bottles to produce these marvellous condiments. Bruise or crush the herbs to increase their surface area for maximum flavour extraction during steeping. Complete directions are given in the recipes. These vinegars keep for a considerable time refrigerated, although the flavours do decline as time goes by; optimum storage time is about 6 months. A sprig of fresh herb added to the bottle after steeping gives extra eye-appeal. Pink Peppercorn (page 189) and Cider Sage (page 189) are intriguing examples of our specialty vinegars.

Ways to Use Flavoured Vinegars

- Use in marinades to tenderize and flavour meats
- Substitute for plain vinegar in salad dressing or vinaigrette recipes
- Splash over steamed or stir-fried vegetables as a salt and fat replacement
- Substitute for lemon juice (except in sweet spread recipes)
- Toss a fruit vinegar and a bit of sugar with fresh berries
- Use to marinate cucumber, tomato and carrot slices
- Combine with low-fat yogurt, mayonnaise or sour cream to make a dip
- Add a splash of herb vinegar to liven up hot or cold soups, sauces and stews.
- Blend equal parts of honey and fruit vinegar to make a dressing for fresh fruit
- Add a dash of herb vinegar to tomato juice or a Bloody Mary

Oils

The multitude of flavoured oils available in stores encouraged us to add a selection of flavoured oil recipes. Stocking a pantry with a similar selection of purchased ones would be expensive, and they are so easily made at home. Furthermore, flavoured oils offer several pluses in today's world of lower-fat cooking. These oils provide interesting flavour to otherwise bland foods, and you can use less of a flavoured oil than a plain one because they deliver such fantastic flavour. In fact, these oils are so highly flavoured that we like to make them in small batches since a little goes such a long way.

Flavoured oils are made by infusing foods such as garlic, herbs or chile peppers into an oil. A bland, pale oil is frequently used to prevent masking the added flavours and colours. Canola oil is ideal. It is fairly tasteless and colourless. As well, it is inexpensive and is the lowest of all oils in unsaturated fats. However, you may prefer to use extra virgin olive oil for the flavour that only olive oil provides.

The problem with infusing oils with fresh foods is *Clostridium botulinum* spores. Although found widely on foods, the spores are seldom a concern because they find few conditions where they can grow. However, whenever a fresh food, such as garlic, is immersed in oil and kept at temperatures over 50°F (10°C), the spores can grow and produce a potentially fatal toxin.

Homemade flavoured oils must be prepared and stored as directed to avoid any danger of botulism. Be sure to follow the recipe instructions closely and you won't need to be concerned. All our flavoured oil recipes follow recommendations made by the United States Food and Drug Administration and confirmed by Health Canada. They suggest the safest and easiest way to infuse an oil with another flavouring is to heat the prepared oil at a low oven temperature (300°F/ 150°C) for a specified time. The mixture is then allowed to cool before it is strained and bottled.

After proper processing, flavoured oils need to kept refrigerated for safety and to prolong their flavour. Oils stored in the refrigerator do become cloudy, but the cloudiness disappears when they return to room temperature.

Serving Suggestions

Use specialty vinegars and flavoured oils to perk up a traditional tossed green salad, a pasta salad or a grain salad such as tabbouleh. In a poultry or meat marinade they do wonders. A splash of a specialty vinegar or a flavoured oil (or both) added to vegetables such as green beans, cauliflower or broccoli is a new eating adventure. Try Garlic Oil (page 190) drizzled on thickly sliced French or Italian breads and broiled—these snacks are just the greatest. And whoever heard of a pizza without a dash of Hot Pepper Oil (page 192)?

List of Recipes

Flavoured Fruit Vinegars

Vinegars are a fundamental part of any good cook's pantry. And yet many are very expensive. Share our enjoyment of these economical, easy-to-prepare homemade vinegars. Most can be made within a week, but allow longer for further flavour development. All add sparkle to foods.

Fruit Vinegar

2 cups	sliced hulled strawberries OR 1 cup (250 mL) raspberries, cherries or blueberries	500 mL
½ cup	rice vinegar	125 mL
2 tsp	granulated sugar	10 mL

1. Place fruit in a clean jar. Heat vinegar to boiling, pour over fruit, cover and steep for several days at room temperature, out of direct sunlight.
2. Strain through a fine sieve, pressing to extract liquid; discard pulp. Heat vinegar with sugar until sugar dissolves; pour into a clean jar with a tight-fitting lid. Store in the refrigerator.

Makes 1 cup (250 mL).

Variation:

Mint leaves or tarragon may be added during steeping.

Citron Vinegar

1	lime	1
½	orange	½
1	lemon, thinly sliced	1
2 cups	white wine vinegar	500 mL
Pinch	each salt and paprika	Pinch
2 tsp	granulated sugar	10 mL
	Strips of lemon rind (optional)	

1. Finely grate rind of lime and orange; combine with lemon slices in a saucepan. Squeeze juice from lime and orange; set aside.
2. Add vinegar, salt and paprika to saucepan; bring to a boil over high heat, remove from heat and let cool. Stir in reserved juice; pour into a clean jar. Cover and steep in a sunny location for 1 week or longer.
3. Strain through a fine sieve and discard pulp. Heat vinegar with sugar until sugar dissolves; pour into a clean jar with a tight-fitting lid. Add lemon rind (if using). Store in the refrigerator.

Makes about 1½ cups (375 mL).

Creamy Citron Dressing
One of the neatest ways to use this vinegar is in an oil and vinegar dressing with crumbled feta cheese.
Place 2 tbsp (25 mL) crumbled feta cheese, 2 tbsp (25 mL) olive oil, 2 tbsp (25 mL) water, 1 tbsp (15 mL) Citron Vinegar and 1 tbsp (15 mL) mayonnaise in a blender or food processor; blend until smooth.
Makes ½ cup (125 mL).

Chile Lime Vinegar with Cilantro

Perfect for mixing into a rice or fresh vegetable salad.

	rind of 1 lime	
1 cup	white wine vinegar	250 mL
2	dry hot chile peppers	2
2	sprigs fresh cilantro	2

1. Place lime rind, vinegar and chiles in a small saucepan. Bring to a boil over high heat; reduce heat, cover and boil gently for 10 minutes. Pour into a clean jar and let cool.
2. Crush or bruise cilantro. Add cilantro to vinegar, cover and steep in a cool, dark place for several weeks.
3. When flavour is satisfactory, strain vinegar and discard solids. Pour into a clean jar with tight-fitting lid. Store in the refrigerator.

Makes 1 cup (250 mL).

Flavoured Herb Vinegars

Experiment with various herbs to make flavourful vinegars. Crush or bruise the herbs before adding them for best release of flavour.

Basic Fresh Herb Vinegar

2 cups	vinegar (white wine, red wine, rice, white or cider)	500 mL
½ cup	fresh herbs (rosemary, sage, tarragon, thyme, basil, parsley, chives and chive blossoms, mint, dill, oregano) OR 3 tbsp (45 mL) dried	125 mL

1. Bring vinegar to a boil.
2. Crush or bruise fresh herbs. Place herbs in a clean jar and pour in vinegar. Cover and steep in a sunny location for 2 weeks or longer.
3. Taste vinegar occasionally and when flavour is satisfactory, strain vinegar and pour into a clean jar with a tight-fitting lid. Add a fresh sprig of herb to the jar if desired. Store in the refrigerator.

Makes 2 cups (500 mL).

Suggested combinations:
- Red wine vinegar with sage, oregano or thyme
- White wine vinegar with chive blossoms, basil or parsley
- Rice vinegar with rosemary
- Cider vinegar with tarragon, dill or mint
- White wine vinegar with lemon herbs (lemon balm, lemon basil, lemon thyme or lemon verbena)

Honey Herb Vinegar

The honey gives the vinegar much the same sweetness of balsamic vinegar.

2 cups	red wine vinegar	500 mL
2 tbsp	honey	25 mL
½ cup	fresh thyme or basil leaves	125 mL

1. Bring vinegar to a boil; stir in honey until dissolved.
2. Crush or bruise herbs. Place herbs in a clean jar and pour in vinegar. Cover and steep in a sunny location for 2 weeks or longer, tasting occasionally.
3. When flavour is satisfactory, strain vinegar and pour into a clean jar with a tight-fitting lid. Store in the refrigerator.

Makes 2 cups (500 mL).

Red Wine and Rosemary Vinegar

A pungent, colourful vinegar to use for salad dressings or marinades.

2 cups	red wine vinegar	500 mL
1 tsp	granulated sugar	5 mL
½ cup	fresh rosemary leaves	125 mL
½	clove garlic, crushed	½

1. Bring vinegar and sugar to a boil.
2. Crush or bruise rosemary. Place rosemary and garlic in a clean jar and pour in vinegar. Cover and steep in a sunny location for up to 2 weeks, tasting occasionally.
3. When flavour is satisfactory, strain vinegar and pour into a clean jar with a tight-fitting lid. Store in the refrigerator.

Makes 2 cups (500 mL).

Variation:

Cranberry and Rosemary Vinegar: Use white vinegar, omit garlic and add ¼ cup fresh cranberries to jar before pouring in vinegar.

Garlic Vinegars

Garlic vinegars are among our favourites to use for salad dressings and the many dishes where vinegar is added.

Red Wine Oregano Garlic Vinegar

2 cups	red wine vinegar	500 mL
1 cup	fresh oregano	250 mL
3	cloves garlic, crushed	3

1. Bring vinegar to a boil. Crush or bruise oregano. Place oregano and garlic in a clean jar.
2. Pour vinegar into jar, cover and steep in a sunny location for up to 2 weeks, tasting occasionally.
3. When flavour is satisfactory, strain vinegar and pour into a clean jar with a tight-fitting lid. Store in the refrigerator.

Makes 2 cups (500 mL).

Chile Herb Garlic Vinegar

2 cups	white wine vinegar	500 mL
3–4	sprigs fresh herb such as basil or thyme	3–4
1	clove garlic, crushed	1
1	hot chile pepper	1

1. Bring vinegar to a boil. Crush or bruise herb. Place herb, garlic and chile in a clean jar.
2. Proceed as in steps 2 and 3 above.

Makes 2 cups (500 mL).

Other Specialty Vinegars

Provence-Style Vinegar

To retain the flavour after bottling this delicate vinegar, add 2 fresh strips of orange rind, 1 thyme sprig, 1 slice shallot, 1 bay leaf and a few fennel seeds.

2 cups	white wine vinegar	500 mL
½ cup	fresh thyme leaves	125 mL
5	wide strips orange rind	5
⅓ cup	thinly sliced shallots	75 mL
2	bay leaves	2
2 tsp	fennel seeds	10 mL

1. Bring vinegar to a boil. Crush or bruise thyme. Place thyme, orange rind, shallots, bay leaves and fennel seeds in a clean jar.
2. Pour vinegar into jar, cover and steep in a cool, dark place for several weeks.
3. Taste vinegar occasionally and when flavour is satisfactory, strain vinegar and pour into a clean jar with a tight-fitting lid. Store in the refrigerator.

Makes 2 cups (500 mL).

Sherry Vinegar

Sherry vinegar is a superb source of flavour to enliven many foods. Try adding it to hot or cold soups or a simple vinaigrette.

1 cup	white wine vinegar	250 mL
1 cup	sherry	250 mL

1. Bring vinegar and sherry just to a boil, cool slightly and pour into a clean jar with a tight-fitting lid.
2. Steep in a cool, dark place for several weeks. Then store in the refrigerator.

Makes 2 cups (500 mL).

For squares, cakes and other
delightful recipes using preserves,
see pages 213 to 230.

Cider Sage Vinegar

Sage gives a special flavour to this vinegar; use only the tender stems, discarding the woody ones. Use this vinegar as a glaze for ham or pork. It is also interesting mixed with oil to make a vinaigrette.

2 cups	cider vinegar	500 mL
1	bunch fresh sage leaves	1
3	cinnamon sticks	3
½ tsp	whole allspice	2 mL
¼ tsp	whole cloves	1 mL

1. Bring vinegar to a boil. Crush or bruise sage. Half fill a clean pint (500 mL) jar with sage leaves. Add cinnamon sticks, allspice and cloves.
2. Pour vinegar into jar, cover and steep in a sunny location for up to 2 weeks, tasting occasionally.
3. When flavour is satisfactory, strain vinegar and pour into a clean jar with a tight-fitting lid. Add some of the sage leaves and spices, if desired. Store in the refrigerator.

Makes 2 cups (500 mL).

Pink Peppercorn Vinegar

The pink peppercorns give a beautiful colour to this simple-to-make vinegar. Use both green and pink for a variation.

2 cups	white wine vinegar	500 mL
2 tbsp	pink peppercorns	25 mL

1. Bring vinegar and peppercorns to a boil, reduce heat and boil gently for 5 minutes. Pour into a clean jar with a tight-fitting lid.
2. Steep in a cool, dark place for several weeks. Then store in the refrigerator.

Makes 2 cups (500 mL).

Variations:

Pink and Green Peppercorn Vinegar: Use 1 tbsp (15 mL) green and 1 tbsp (15 mL) pink peppercorns.

Tarragon Green Peppercorn Vinegar: Add ½ cup (125 mL) fresh tarragon or 2 tbsp (25 mL) dried.

Creative ideas to package your preserves
. can be found on page 231.

Flavoured Oils

Flavoured oils are enjoying increasing popularity because they enliven a wide variety of foods. Just a few drops add great flavour, making them an excellent way to lower the fat in our diets. Many are the recipes for these oils, but they do not always make a safe product. Be sure to read the introduction to this section, carefully follow the instructions given in the recipes and see "Food Safety Alert" on page 191. Always store the finished oils in the refrigerator and use within a month.

Garlic Oil

Keep Garlic Oil in your refrigerator to use when sautéing vegetables for a stir-fry or to add flavour to a pilaf.

1 cup	extra virgin olive or canola oil	250 mL
6	large cloves garlic, halved lengthwise	6

1. Wash and dry a 28 oz (796 mL) can with lid removed. Place oil and garlic in can and set on a baking sheet. Bake in a 300°F (150°C) oven for 1 hour or until the garlic cloves are medium brown. Remove can to a rack to cool for 30 minutes.
2. Line a small strainer with a coffee filter or several layers of cheesecloth. Strain oil into a clean glass jar, cover and store in the refrigerator. Use within a month.

Makes about 1 cup (250 mL).

Traditional Vinaigrette: Combine ¼ cup (50 mL) Garlic Oil or other flavoured oil, 2 tbsp (25 mL) balsamic, red wine or white wine vinegar, 2 tbsp (25 mL) water and 1 pinch each granulated sugar, salt and freshly ground pepper in a small jar. Cover and shake well. Refrigerate until ready to use.
Makes ½ cup (125 mL).

Herbed Pepper Garlic Oil

We like to add a few drops of this flavourful oil to fresh green beans or other steamed vegetables. A little drop is all that is needed.

1 cup	canola or extra virgin olive oil	250 mL
1 tbsp	chopped fresh rosemary, basil or oregano	15 mL
1	hot chile pepper, sliced	1
2	large cloves garlic, halved lengthwise	2

1. Wash and dry a 28 oz (796 mL) can with lid removed. Place oil, rosemary, chile pepper and garlic in can and set on a baking sheet. Bake in a 300°F (150°C) oven for 1 hour or until the chile pepper and garlic cloves are medium brown. Remove can to a rack to cool for 30 minutes.
2. Line a small strainer with a coffee filter or several layers of cheesecloth. Strain oil into a clean glass jar, cover and store in the refrigerator. Use within a month.

Makes 1 cup (250 mL).

Variation:

Herbed Garlic Oil
For an excellent herbed garlic oil, simply eliminate the hot chile pepper.

Food Safety Alert: The amount of oil and the size and shape of the container used to heat flavoured oils are essential to making an oil that will not support growth of harmful microorganisms. If you want a larger quantity of oil than that from one recipe, put a second batch of ingredients (a second recipe) into a separate can; two cans can be heated at the same time in the oven. Don't put more than one cup (250 mL) of oil in one can or use a smaller metal container than the 28 oz (796 mL) can. When the oil has cooled, it is of the utmost importance to keep the oil refrigerated when it's not in use and to keep it no longer than one month. It is a good idea to verify the temperature in your oven with an oven thermometer. Also read the introduction to this chapter for more about avoiding botulism poisoning.

Hot Pepper Oil

Choose a hot pepper from the Chile Pepper Heat Scale chart on page 133. One Scotch bonnet gives as hot an oil as we've tasted. Use it to enliven pizzas or other spicy food. Add several dried peppers and a dry bay leaf to finished oil for an attractive gift.

1 cup	extra virgin olive or canola oil	250 mL
1–2	hot chile peppers OR 8 dried red chiles	1–2
2	bay leaves	2
6	peppercorns	6

1. Wash and dry a 28 oz (796 mL) can with lid removed. Place oil, chile peppers, bay leaves and peppercorns in can and set on a baking sheet. Bake in a 300°F (150°C) oven for 1 hour or until the chiles are medium brown. Remove can to a rack to cool for 30 minutes.
2. Line a small strainer with a coffee filter or several layers of cheesecloth. Strain oil into a clean glass jar, cover and store in the refrigerator. Use within a month.

Makes 1 cup (250 mL).

Herbed Orange Oil

Citrus oils make a great base for a fruit salad dressing.

1 cup	canola or extra virgin olive oil	250 mL
2	strips orange rind	2
1	sprig fresh rosemary	1
6	whole black peppercorns	6

1. Wash and dry a 28 oz (796 mL) can with lid removed. Place oil, orange rind, rosemary and peppercorns in can and set on a baking sheet. Bake in a 300°F (150°C) oven for 1 hour or until the orange rind is medium brown. Remove can to a rack to cool for 30 minutes.
2. Line a small strainer with a coffee filter or several layers of cheesecloth. Strain oil into a clean glass jar, cover and store in the refrigerator. Use within a month.

Makes 1 cup (250 mL).

Lemon Thyme Oil

The next time you barbecue chicken, add some finely chopped onions to this oil to brush on during cooking.

1 cup	extra virgin olive or canola oil	250 mL
2	strips lemon rind	2
4	sprigs fresh thyme or sage	4

1. Wash and dry a 28 oz (796 mL) can with lid removed. Place oil, lemon rind, and thyme in can and set on a baking sheet. Bake in a 300°F (150°C) oven for 1 hour or until the lemon rind is medium brown. Remove can to a rack to cool for 30 minutes.
2. Line a small strainer with a coffee filter or several layers of cheesecloth. Strain oil into a clean glass jar, cover and store in the refrigerator. Use within a month.

Makes 1 cup (250 mL).

Gingered Thai Oil

Sauté onions in this oil to give a Thai influence to stir-fries, or stir a few drops into steamed rice.

1 cup	canola or extra virgin olive oil	250 mL
2	strips lemon rind	2
1	1-inch (2.5 cm) piece gingerroot, cut up	1
1	Scotch bonnet or other hot chile pepper, cut up	1
1	large clove garlic, halved lengthwise	1

1. Wash and dry a 28 oz (796 mL) can with lid removed. Place oil, lemon rind, gingerroot, chile pepper and garlic in can and set on a baking sheet. Bake in a 300°F (150°C) oven for 1 hour or until the lemon rind, chile pepper and garlic are medium brown. Remove can to a rack to cool for 30 minutes.
2. Line a small strainer with a coffee filter or several layers of cheesecloth. Strain oil into a clean glass jar, cover and store in the refrigerator. Use within a month.

Makes 1 cup (250 mL).

Chapter Twelve

The Finishing Touch

W HENEVER we think the sweet thoughts of dessert, "visions of sugar plums dance in our heads." And that's how we approached this chapter.

Not all desserts can have a lid on them, but those that can, make fast-finish desserts. Some of ours are complete desserts made earlier to allow flavours to develop over a number of days. Apple Cranberry Mincemeat (page 202) is a perfect example, as are the many combinations of fresh fruits with liqueurs from our chart of Fruit with Spirit (page 196). Others, like the Brandied Dried Fruit Preserves (page 200) served over ice cream, an angel food cake or a layer cake, make a sublime ending to a meal. Once preserved in a boiling-water canner, these jars will wait on your pantry shelves for just that right moment—likely a day when too many other activities come in the way of dessert preparation.

Having dessert sauces and syrups on hand in a jar opens up endless possibilities for making more complex recipes easier. Think of using a dessert sauce or syrup as part of the liquid in a cake batter. Or using Cherry Compote (page 198) to make cherries jubilee. And a real favourite, so elegant yet very simple, is fresh pears and chèvre cheese in Rosemary Wine Syrup (page 206) served with a sweet wafer cookie.

People have been making homemade liqueurs for years from kits, but less frequently from fresh fruits. Naturally, the fresh fruit ones taste the most natural. Try one of our fruit liqueurs for your next end-of-meal digestive.

We hope you will agree that "putting a lid on" these creations is the perfect finishing touch.

List of Recipes

Fruit with Spirit

Fruit can be magically transformed into an enchanting dessert with the addition of your favourite wine or liqueur.

Fruit and Preparation for 4 pint (500 mL) jars

Fruit	Preparation of Fruit	Spirit per Pint (500 mL) Jar	Processing Time
Apricots 7 cups (1.75 L) halved (about 2 lb/1 kg)	Dip in boiling water 30–60 seconds to blanch. Remove skins, cut in half and pit.	1 tbsp (15 mL) peach schnapps or Amaretto OR 2 tbsp (25 mL) brandy or rum	20 minutes
Cherries (sweet) 7 cups (1.75 L) (about 2 lb/1 kg)	Pit.	1 tbsp (15 mL) kirsch or Amaretto OR 2 tbsp (25 mL) port or sherry	15 minutes
Grapes, seedless 8 cups (2 L) whole or halved (about 2 lb/1 kg)	Remove stems. Leave whole or cut in half.	1 tbsp (15 mL) orange liqueur or kirsch OR 2 tbsp (25 mL) port or sherry	15 minutes
Peaches or Nectarines 6 cups (1.5 L) sliced (about 2.5 lb/1.2 kg)	Dip peaches in boiling water 30–60 seconds to blanch. Slip off skins, cut in half, pit and slice.	1 tbsp (15 mL) orange liqueur or Chambord OR 2 tbsp (25 mL) peach schnapps	20 minutes
Pears 6 cups (1.5 L) sliced (about 2.5 lb/1.2 kg)	Peel, core and slice.	2 tsp (10 mL) crème de menthe OR 1 tbsp (15 mL) Frangelica OR 2 tbsp (25 mL) brandy or rum	20 minutes
Pineapple 1 whole, cubed	Peel deep enough to remove eyes. Cut into quarters and remove core. Cut into cubes or spears.	2 tsp (10 mL) crème de menthe or kirsch OR 2 tbsp (25 mL) brandy or rum	15 minutes
Plums 6 cups (1.5 L) sliced (about 2.5 lb/1.2 kg)	Pit and slice.	1 tbsp (15 mL) Amaretto or peach schnapps OR 2 tbsp (25 mL) port or rum	15 minutes

Syrup for 4 pint (500 mL) jars:

1	lemon	1
2½ cups	water	625 mL
2 cups	granulated sugar	500 mL

Canning Procedure:

1. Partially fill a boiling-water canner with hot water. Place clean mason pint (500 mL) jars in canner, cover and begin to bring water to a boil over high heat.
2. Prepare fruit as directed. As you work, place all fruits except grapes, pineapple and sweet cherries in 4 cups (1 L) water mixed with ¼ cup (50 mL) lemon juice. Drain fruit before placing in jars.
3. Prepare syrup: Grate rind of lemon; place in a medium saucepan. Squeeze 2 tbsp (25 mL) juice from lemon; add to rind. Stir in water and sugar. Bring to a boil over high heat and boil gently for 1 minute. Meanwhile, place snap lids in hot or boiling water according to manufacturer's directions.
4. Remove jars from canner. Add choice of liqueur as suggested in chart and pack fruit into jars. Ladle hot syrup over fruit to within ½ inch (1 cm) of rim (head space). Remove air bubbles by sliding a small clean wooden or plastic spatula between glass and food; readjust head space to ½ inch (1 cm). Wipe jar rim to remove any stickiness. Centre snap lid on jar; apply screw band just until fingertip tight. Place jars in canner and adjust water level to cover jars by 1–2 inches (2.5–5 cm). Cover canner and return water to a boil. Process for times given in chart.
5. Remove jars from canner and cool for 24 hours. Check jar seals (sealed lids turn downward). Wipe jars, label and store in a cool, dark place.

Sugar greatly enhances the flavour and texture of canned fruit, but it is not essential in the preserving process. Reduce the sugar if you choose.

Cherry Compote

Compotes are fresh or dried fruits slowly cooked in a sugar syrup to retain their shape. Compotes frequently contain wine or liqueur. This one is delicious served over a slice of cheesecake or pound cake.

1 cup	dry red wine	250 mL
⅔ cup	granulated sugar	150 mL
2 tsp	lemon juice	10 mL
4 cups	fresh or frozen pitted sour cherries	1 L
1 tbsp	cornstarch	15 mL
1 tbsp	water	15 mL
1 tbsp	kirsch	15 mL

1. Place wine, sugar and lemon juice in a large saucepan. Bring to a boil over high heat, stirring to dissolve sugar. Add cherries; return to a boil. Reduce heat and boil gently, uncovered, for 15 minutes, stirring occasionally.
2. Remove hot jars from canner. Remove cherries from liquid with a slotted spoon; pack into the jars. Continue simmering syrup until it is reduced to ⅔ cup (150 mL).
3. Stir together cornstarch and water; stir into syrup. Return to a boil and boil gently for 1 minute, stirring constantly. Remove from heat and stir in kirsch.
4. Pour syrup over cherries to within ½ inch (1 cm) of rim (head space). Process for 15 minutes for half-pint (250 mL) jars and 20 minutes for pint (500 mL) jars as directed on page 98 (Longer Time Processing Procedure).

Makes 3 cups (750 mL).

Cherries Jubilee
Warm 2 tbsp (25 mL) brandy and pour over Cherry Compote. Set alight with a match. Serve over vanilla ice cream. For easy warming, place brandy in small micro-wavable container and microwave, uncovered, at High (100%) for 10 seconds.

Hazelnut Blueberry Mango Sauce

Enjoy this blending of the tropical flavour of mango with down-home blueberry. Serve it over pancakes, waffles or angel food cake. If mangoes are unavailable, substitute nectarines or peeled peaches.

2	mangoes, peeled and diced	2
2 cups	fresh or frozen unsweetened blueberries	500 mL
1 cup	granulated sugar	250 mL
1 cup	apple juice or water	250 mL
2 tbsp	lemon juice	25 mL
1 tsp	grated lemon rind	5 mL
2 tbsp	hazelnut liqueur OR 1½ tsp (7 mL) almond extract	25 mL

1. Combine mangoes, blueberries, sugar, apple juice, lemon juice and lemon rind in a medium stainless steel or enamel saucepan. Bring to a boil, reduce heat and simmer, uncovered, for 25 minutes or until fruit is softened and liquid has thickened. Stir in liqueur and cook for 5 minutes.
2. Remove hot jars from canner and ladle compote into jars to within ½ inch (1 cm) of rim (head space). Process for 10 minutes for half-pint (250 mL) jars and 15 minutes for pint (500 mL) jars as directed on page 98 (Longer Time Processing Procedure).

Makes 4 cups (1 L).

Variation:

8 nectarines or peeled peaches may be substituted for the mangoes.

Brandied Dried Fruit Preserves

Elegant desserts are easy and fast with this spirited preserve on hand. Serve over ice cream, frozen yogurt, fresh fruits and cake. You will probably find other uses as well.

1	pkg (227 g) dried mixed fruits (about 1½ cups/375 mL if you want to make your own selection)	1
1	1-inch (2.5 cm) piece crystallized ginger, thinly sliced	1
⅓ cup	lightly packed brown sugar	75 mL
2	thin strips orange rind, chopped	2
½ cup	brandy or cognac	125 mL

1. Place dried fruits and ginger in a saucepan. Cover with cold water, bring to a boil, remove from heat and let cool. Refrigerate for 8 hours or overnight.
2. Drain fruit, reserving liquid. Add enough water to liquid to make ¾ cup (175 mL). In a saucepan, combine liquid, sugar and orange rind. Bring to a boil, stirring, until sugar has dissolved. Add drained fruit; warm. Remove from heat and stir in brandy.
3. Remove hot jars from canner. Pack fruit into jars. Ladle liquid over fruit in jars to within ½ inch (1 cm) of rim (head space). Process for 10 minutes for half-pint (250 mL) jars and 15 minutes for pint (500 mL) jars as directed on page 98 (Longer Time Processing Procedure).

Makes 2 cups (500 mL).

Spiced Cranberry Preserves

Add making this preserve to your "must do" list. Kept on hand, it provides a fast way to brighten up ice cream, frozen yogurt, cake or rice pudding.

1	pkg (340 g) fresh or frozen cranberries (about 4 cups/1 L)	1
1	large apple, peeled, cored and diced	1
1	large pear, peeled, cored and diced	1
½ cup	golden raisins	125 mL
1 cup	granulated sugar	250 mL
1 tbsp	grated orange rind	15 mL
½ cup	orange juice	125 mL
1 tsp	ground cinnamon	5 mL
¼ tsp	ground nutmeg	1 mL
⅓ cup	orange liqueur	75 mL

1. Combine cranberries, apple, pear, raisins, sugar, orange rind, orange juice, cinnamon and nutmeg in a medium saucepan. Bring to a boil over high heat, reduce heat and boil gently, uncovered, for 20 minutes or until thickened, stirring frequently. Remove from heat and stir in liqueur.
2. Remove hot jars from canner and ladle preserves into jars to within ½ inch (1 cm) of rim (head space). Process for 10 minutes for half-pint (250 mL) jars and 15 minutes for pint (500 mL) jars as directed on page 98 (Longer Time Processing Procedure).

Makes 4 cups (1 L).

Variation:

Raspberry Cranberry Dessert Preserves
Replace pear and apple with 1½ cups (375 mL) fresh or frozen unsweetened raspberries.

Apple Cranberry Mincemeat

Our favourite mincemeat because of its fruitiness and freshness, this one is positively addictive. In fact, it is good enough to eat right out of the jar. Use it in pies and squares or warm it as an ice cream topping.

6 cups	finely chopped tart apples (2 lb/1 kg)	1.5 L
1 cup	sultana raisins	250 mL
1 cup	Thompson raisins	250 mL
1 cup	currants	250 mL
⅔ cup	brown sugar	150 mL
½ cup	dried cranberries	125 mL
⅓ cup	water	75 mL
	Grated rind and juice of 2 lemons	
2 tsp	ground cinnamon	10 mL
½ tsp	ground cloves	2 mL
¼ cup	brandy	50 mL

1. Combine apples, sultana raisins, Thompson raisins, currants, sugar, cranberries, water, lemon rind, lemon juice, cinnamon and cloves in a large stainless steel or enamel saucepan. Bring to a boil over high heat, reduce heat, cover and simmer for 20 minutes, stirring frequently. Stir in brandy and return to a boil.
2. Remove hot jars from canner and ladle preserves into jars to within ½ inch (1 cm) of rim (head space). Process for 20 minutes for half-pint (250 mL) jars and pint (500 mL) jars as directed on page 98 (Longer Time Processing Procedure).

Makes 6 cups (1.5 L).

Tip: Be sure to try Apple Cranberry Mincemeat Squares (page 226) the next time you need an accompaniment for coffee or tea.

Sweet Fruit Sauces

Keep these fruit sauces on hand in the refrigerator. They are wonderful and fast toppings for puddings, fresh fruit, pancakes, crêpes or ice cream. These small recipes can easily be doubled or tripled.

Blueberry Maple Sauce

2 cups	fresh or frozen unsweetened blueberries	500 mL
¼ cup	water	50 mL
1 tbsp	lemon juice	15 mL
1	cinnamon stick	1
½ cup	granulated sugar	125 mL
½ cup	maple syrup	125 mL

1. Combine blueberries, water, lemon juice and cinnamon stick in a 4-cup (1 L) microwavable bowl. Microwave, uncovered, on High (100%) for 5 minutes, stirring once. Stir in sugar and maple syrup. Microwave, uncovered, on High for 3 minutes or until sugar is dissolved and mixture comes to a boil. Let cool.
2. Discard cinnamon stick and store sauce in a tightly sealed container in refrigerator for up to 1 month.

Makes 2 cups (500 mL).

Warm Apple Cider Sauce

Enjoy this sauce strained and added to hot tea for a warming experience after a winter outing.

2 tbsp	butter or margarine	25 mL
1 cup	apple cider	250 mL
¾ cup	packed brown sugar	175 mL
2 tbsp	cornstarch	25 mL
1–2 tbsp	rum, brandy or Calvados (optional)	15–25 mL

1. Melt butter in a microwavable 2-cup (500 mL) container on High (100%) for 30 seconds. Stir together cider, sugar and cornstarch. Whisk into butter. Microwave, uncovered, on High for 3 minutes or until smooth and thickened. Stir in rum (if using). Serve warm.

Makes 1½ cups (375 mL).

Cranberry Raspberry Sauce

2½ cups	fresh or frozen cranberries	625 mL
1	pkg (300 g) frozen unsweetened raspberries, thawed	1
¾ cup	cranberry fruit cocktail	175 mL
½ cup	lightly packed brown sugar	125 mL
½ cup	corn syrup	125 mL
½ tsp	almond extract	2 mL

1. Combine cranberries, raspberries, cranberry cocktail, sugar and corn syrup in a medium saucepan. Bring to a boil, reduce heat and boil gently for 15 minutes or until fruit is softened. Remove from heat, purée in a food processor or blender until smooth. Stir in almond extract and cool.

Makes 3½ cups (875 mL).

Brandied Fruit Sauce

Looking for the perfect hostess gift? Here's a sensational and simple recipe to fold into whipped cream, serve over ice cream or fresh fruit or use as a pudding topping. The recipe is so easy, why not double it and have some yourself!

1 cup	each golden raisins and mixed candied peel	250 mL
½ cup	candied pineapple cubes, diced	125 mL
⅓ cup	each candied red and green cherries	75 mL
⅓ cup	diced crystallized ginger	75 mL
¼ cup	dried cranberries	50 mL
½ tsp	ground cinnamon	2 mL
Pinch	each ground cloves and allspice	Pinch
1 cup	dark rum or brandy	250 mL

1. Place raisins, peel, pineapple, cherries, ginger and cranberries in a sterilized quart (1 L) jar.
2. Stir together cinnamon, cloves, allspice and rum in a small saucepan. Heat until just warm. Pour over fruit. Seal jar securely and rotate each day for 6 days.
3. Spoon into smaller attractive jars for gift giving. Keeps in a cool, dark place almost indefinitely.

Makes 2 cups (500 mL).

Wine-Berry Sauce

2 cups	strawberries, sliced	500 mL
¼ cup	granulated sugar	50 mL
⅓ cup	dry red wine	75 mL
2 tsp	Amaretto OR ¼ tsp (1 mL) almond extract	10 mL

1. Place strawberries in a bowl and toss with sugar. Let stand for 12 hours.
2. Drain liquid from berries into a small saucepan; add wine. Bring to a boil over high heat, reduce heat and boil gently for 1 minute.
3. Pour wine and liqueur over berries and mix well. Refrigerate for 1 week or freeze for longer storage.

Makes 2 cups (500 mL).

Variation:

Raspberries with Port: Raspberries, port and 1 tbsp (15 mL) brandy in place of strawberries, red wine and liqueur are also delicious served over ice cream.

Dessert Syrups

The Rosemary Wine Syrup is a knockout with fresh fruit and chèvre cheese. Arrange any combination of fresh fruit and cheese on individual dessert plates. Drizzle with the syrup and garnish with fresh mint. Mint Lime Syrup is marvellous over fresh fruit, especially with cubes of cantaloupe and honeydew melon. We love Cinnamon Rum Syrup over cooked rhubarb sauce, ice cream and especially on fruit pancakes. And Blueberry Syrup makes a fabulous dessert over crêpes, waffles, ice cream and frozen yogurt.

Rosemary Wine Syrup

½ cup	granulated sugar	125 mL
½ cup	dry white wine	125 mL
¼ cup	fresh rosemary leaves	50 mL
¼ cup	water	50 mL
2 tbsp	balsamic vinegar	25 mL
½ tsp	peppercorns	2 mL
2	strips lemon rind	2
2	bay leaves	2

1. Combine sugar, wine, rosemary, water, vinegar, peppercorns, lemon rind and bay leaves in a small saucepan. Bring to a boil over high heat, cover, reduce heat and boil gently for 5 minutes. Remove from heat and let cool. Strain mixture through a fine sieve; discard solids.
2. Pour syrup into a clean jar with a tight-fitting lid and store in the refrigerator.

Makes about 1¼ cups (300 mL).

Mint Lime Syrup

1	lime	1
½ cup	granulated sugar	125 mL
½ cup	white wine	125 mL
¼ cup	chopped fresh mint OR 1 tbsp (15 mL) dried	50 mL

1. Remove 2 strips rind from lime. Squeeze 2 tbsp (25 mL) juice from lime; set aside.
2. Combine rind, sugar, wine and mint in a small saucepan. Bring to a boil over high heat, cover, reduce heat and boil gently for 5 minutes. Remove from heat and let cool. Strain mixture through a fine sieve; discard solids.
3. Stir in lime juice. Pour syrup into a clean jar with a tight-fitting lid and store in the refrigerator.

Makes 1 cup (250 mL).

Cinnamon Rum Syrup

¾ cup	granulated sugar	175 mL
½ cup	water	125 mL
1 tsp	ground cinnamon	5 mL
1 tbsp	dark rum	15 mL
1 tsp	lemon juice	5 mL

1. Combine sugar, water and cinnamon in a small saucepan. Bring to a boil over high heat and boil until sugar is dissolved. Remove from heat; stir in rum and lemon juice. Let cool.
2. Pour syrup into a clean jar with a tight-fitting lid and store in the refrigerator.

Makes 1 cup (250 mL).

Blueberry Syrup

2 cups	fresh or frozen blueberries	500 mL
1 cup	granulated sugar	250 mL
⅓ cup	water	75 mL
1 tbsp	lemon juice	15 mL
½ tsp	ground cinnamon	2 mL

1. Combine blueberries, sugar, water, lemon juice and cinnamon in a medium saucepan. Bring to a boil over high heat, reduce heat, cover and boil gently for 10 minutes or until fruit is tender.
2. Strain through a lined sieve; discard solids.
3. Pour syrup into a clean jar with a tight-fitting lid and store in the refrigerator.

Makes 1 cup (250 mL).

Liqueurs

Making liqueurs is within the reach of the home cook. As well as being econom-ical, the liqueurs are typically lighter and less sweet than the commercial ones.

Basic Fruit Liqueur

Try a variety of fruits such as raspberries, strawberries, sour cherries, nectarines, peaches and plums. A splash of a fruit liqueur does wonders for fresh fruit to add elegance to a simple dessert. Added to soda water, it makes a refreshing summer spritzer. And of course pour some over ice cream or frozen yogurt. Either vodka or 40% alcohol may be used for these liqueurs.

2 cups	fruit cut in halves or slices if larger	500 mL
1 cup	alcohol or vodka	250 mL
2 tbsp	brandy	25 mL
½ cup	granulated sugar	125 mL

1. Place fruit in a clean 1-quart (1 L) jar. Stir in alcohol and brandy. Marinate, covered, for several weeks in a cool, dark place.
2. Strain though a fine sieve lined with cheesecloth; reserve fruits for another use. Stir in sugar.
3. Pour into sterilized bottles, cork, label and store in a cool place for at least 4 weeks.

Makes about 2 cups (500 mL).

Citron Liqueur

2	limes	2
1	orange	1
1	lemon	1
1½ cups	alcohol or vodka	375 mL
¾ cup	granulated sugar	175 mL

1. Remove rind from limes, orange and lemon. Place rind in a clean jar with a tight-fitting lid. Add alcohol; cover and let stand at room temperature for 2 weeks.
2. Discard peel; stir in sugar. Cover and let stand at room temperature for 1 week or longer. Shake jar occasionally to help sugar dissolve.

Makes 2½ cups (625 mL).

Cranberry Liqueur

1	can (9.7 oz/275 mL) frozen cranberry juice cocktail, thawed	1
⅔ cup	alcohol or vodka	150 mL
1	vanilla bean	1

Place cranberry juice, alcohol and vanilla bean in a clean jar. Seal and shake until blended. Store in a cool, dark place for at least 2 weeks.

Makes 1¾ cups (425 mL).

Coffee Liqueur

Try it in milk or over crushed ice.

½ cup	freshly brewed espresso	125 mL
½ cup	brown sugar	125 mL
1 cup	alcohol or vodka	250 mL
⅓ cup	brandy or rum	75 mL
1 tsp	vanilla extract	5 mL

1. Place hot coffee in a 2-cup (500 mL) measure and stir in sugar until dissolved.
2. Stir in alcohol, brandy and vanilla. Pour into a clean bottle. Store 4 weeks before using.

Makes 2 cups (500 mL).

Tip: If you wish, ½ cup (125 mL) water and 1 tbsp (15 mL) instant coffee powder may be substituted for the espresso.

Adding Liqueurs

Appetizer Fruit Kebabs: Place cubes of fresh fruit on skewers and marinate for several hours in your favourite liqueur for a nice addition to an appetizer tray.

Frosty Summer Shakes: Add a fruit liqueur to a milkshake made with vanilla ice cream and sweetened fruit.

Angel Dessert: Add strawberry liqueur to sweetened fresh or frozen strawberries. Serve over slices of angel food cake.

Cranberry Chicken: Splash cranberry or raspberry liqueur over grilled chicken.

Orange Baked Pears

Use any fruit liqueur in this recipe for one of the easiest desserts you may ever prepare. Try it another time with plums, apples or nectarines.

Arrange 6 halved peeled pears cut side down in a baking dish. Combine ½ cup (125 mL) orange juice, 3 tbsp (45 mL) fruit liqueur, 2 tbsp (25 mL) liquid honey and 1 tsp (5 mL) grated orange rind. Pour over pears and sprinkle with a pinch of ground cinnamon. Cover and bake in a 350°F (180°C) oven for 15 minutes. Baste pears with liquid. Cover and bake for another 15 minutes or until pears are tender.

Makes 6 servings.

Chapter Thirteen

Let's Open the Lid and Use What's Inside

M ANY OF THE RECIPES in this book have obvious uses. The introductions to each chapter give suggestions on how to serve them and ideas for special uses. Having worked through the recipes in the book and now having a great array of filled jars sitting on our shelves, we began to think of how else to use them. You may find yourself asking this question too. This chapter is a sampling of possibilities. We know you will think of more.

Marmalades, as well as being superb toast toppers, have found their way into squares, fruitcake and tea loaves. Salsas and chutneys are a basis for dips (page 214–217). Pickled Jalapeño Peppers (page 111) found their way into Jalapeño Quesadillas (page 218). Fruit butters turn up in muffins and soups (pages 213 and 219). The list goes on.

We believe that canning and preserving is one of the most creative areas of culinary endeavour. We hope this chapter, along with the rest of the book, will pique your interest. How often we have heard people exclaim, "What, another meal to prepare!" Once a few of the extra "little indulgences" of this book grace your cupboard shelves, "another meal" will become a chance for fun and creativity.

Bon appétit!

List of Recipes

Spiced Plum Butter Bran Muffins

Spiced Plum Butter gives moisture and a very pleasing plum flavour to these hearty and nutritious muffins. The batter for this large-batch recipe will keep for up to 6 days in the refrigerator.

3 cups	all-purpose flour	750 mL
2 cups	natural bran	500 mL
⅔ cup	lightly packed brown sugar	150 mL
1½ tsp	each baking powder, baking soda and ground cinnamon	7 mL
½ tsp	salt	2 mL
Pinch	ground nutmeg	Pinch
½ cup	canola oil	125 mL
2	eggs, lightly beaten	2
1 cup	Spiced Plum Butter (page 80)	250 mL
1 cup	buttermilk	250 mL
¾ cup	raisins	175 mL
⅓ cup	molasses	75 mL
1	medium apple, peeled, cored and chopped	1

1. Stir together flour, bran, sugar, baking powder, baking soda, cinnamon, salt and nutmeg in a medium bowl.
2. Stir together oil, eggs, Spiced Plum Butter, buttermilk, raisins, molasses and chopped apple in a second bowl.
3. Stir liquid into dry ingredients just until moistened; do not overmix.
   To bake Muffins:
4. Pour batter into a sealed container and refrigerator up to 6 days, until ready to use. Spoon batter into lightly greased nonstick or paper-lined medium muffin tins. Bake in 375°F (190°C) oven for 18 minutes or until muffins are lightly browned and firm to the touch. Cool on a wire rack.

Makes 30 medium muffins.

Variation:

Sweet and Chunky Apple Butter (page 79) may replace Spiced Plum Butter.

Appealing Appetizers

When you open the lid of the condiment jar, much of the work is already done for these great appetizers.

Jalapeño Cheddar Canapés

Stir together 1 cup (250 mL) shredded old Cheddar cheese, 1 large egg and 2–3 tsp (10–15 mL) chopped drained Pickled Jalapeño Peppers (page 111). Mound a small spoonful onto sixteen 2-inch (5 cm) toast rounds. Place on a baking sheet and broil until puffed and golden. Makes 16 canapés.

Hot Melted Brie with Chutney

Place whole Brie (approximately 4 inches/10 cm in diameter) in a shallow heat-proof dish. Top with ½ cup (125 mL) chutney (chapter 9) and ¼ cup (50 mL) chopped almonds. Bake in 425°F (220°C) oven for 7 minutes or until cheese softens and starts to melt. Serve warm with apple or pear slices and crackers.

Pesto Pita Pizzas

Cut 2 whole wheat pitas in half. Split each half and spread with 1 tsp (5 mL) Presto Pesto (page 176). Sprinkle each with some chopped sweet red pepper and shredded mozzarella cheese. Bake in 400°F (200°C) oven for 5 minutes or until cheese melts. Cut each pita quarter in half for appetizers or leave uncut for lunch. Makes 16 appetizers or 2–4 lunch servings.

Meatballs with Hot and Spicy Sauce

Prepare your favourite meatball recipe or, to keep this appetizer really simple, buy frozen ones. Place frozen meatballs on a baking sheet. Bake in 400°F (200°C) oven for 20 minutes or until browned and cooked through. Serve with Picante Tomato Sauce (page 174) for dipping. A dollop of horseradish adds extra zip!

Meatballs with Oriental Chutney Sauce

Place ⅓ cup (75 mL) chutney (chapter 9), ½ cup (125 mL) orange juice, 2 tbsp (25 mL) each cornstarch and soy sauce, and 1 tsp (5 mL) mustard in a small saucepan. Simmer about 5 minutes or until smooth and thickened, stirring occasionally. Use as a meatball dipping sauce.

Thai Chicken Wings

Pour ½ cup (125 mL) Chili Thai Sauce (page 172) over 1 lb (500 g) chicken wings or cubed chicken in a baking dish. Bake in 400°F (200°C) oven for 25 minutes, stirring occasionally.

Mustard Appetizers

Spread slices of French bread with Honey Mustard Sauce (page 163). Sprinkle lightly with grated Parmesan cheese and broil until bubbly. Serve warm.

Curried Mango Chutney Dip

Blend ½ cup (125 mL) light mayonnaise, 2 tbsp (25 mL) Mango Chutney (page 154), 1 tbsp (15 mL) lemon juice, 1 chopped green onion and 1 tsp (5 ml) curry powder. Makes ¾ cup (175 mL). This is a must with raw vegetables.

Tantalizing Dips and Spreads

Picante Cream Cheese Dip

Process 1 pkg (250 g) light cream cheese, ½ cup (125 mL) drained Pickled Jalapeño Peppers (page 111), ½ sweet red pepper, chopped, and 1 clove garlic in a food processor until coarsely chopped. Transfer to a bowl and stir in 2 tbsp (25 mL) chopped fresh cilantro. Makes about 1½ cups (375 mL) dip. Extra may be frozen.

Pesto Torta Appetizer

Blend 1 pkg (250 g) softened cream cheese with ½ cup (125 mL) softened butter or margarine. Line a small bowl with plastic wrap. Layer one-third cheese mixture, ¼ cup (50 mL) Presto Pesto (page 176), one-third cheese, ¼ cup (50 mL) pesto and one-third cheese. Cover and chill until firm. Unmould and serve at room temperature with assorted crackers or thinly sliced baguette. Makes 1½ cups (375 mL). This torta can be frozen.

Cranberry Port Conserve Cheese Spread

Blend 1 pkg (250 g) softened light cream cheese, ½ cup (125 mL) Cranberry Port Conserve (page 78) or Cranberry Salsa (page 139) and 2 tsp (10 mL) grated orange rind. Makes 1 cup (250 mL).

Salsa Savvy

Salsa with Mozzarella Toasts

36	thin slices baguette	36
	Softened butter	
1 cup	shredded mozzarella cheese	250 mL
1 cup	Winter Salsa (page 137)	250 mL

1. Place bread slices on a baking sheet. Broil bread until lightly toasted on both sides. Spread one side lightly with butter. Sprinkle with cheese. Freeze if desired.
2. Bake in 375°F (190°C) oven for 5 minutes or just until cheese melts. Serve warm with salsa.

Makes 36 appetizers.

Warm Salsa Cheese Dip

Serve warm with tortilla chips, raw vegetables or crackers.
Combine 1 cup (250 mL) mild or medium salsa (chapter 8), 2 pkg (250 g) cream cheese and 1 cup (250 mL) shredded Monterey Jack cheese in a microwavable bowl. Microwave, uncovered, on Medium (50%) for 4 minutes or until cheese is melted, stirring once. Stir in 1 cup (250 mL) beer.

Makes 1½ cups (375 mL).

Fiesta Nacho Appetizer

An attractive moulded appetizer sure to be a highlight at your party.
Blend 1 cup (250 mL) cottage cheese, 1 pkg (250 g) light cream cheese and ½ cup (125 mL) salsa (chapter 8) in a food processor until smooth. Line a 2-cup (500 mL) bowl with plastic wrap. Pack mixture into bowl, cover and chill for 2 hours or overnight. Unmould cheese onto large serving plate and garnish with shredded lettuce and Cheddar cheese, chopped tomato and green onion. Serve with corn chips for dipping.

Makes 2 cups (500 mL).

Jalapeño Quesadillas

Pickled Jalapeño Peppers are the start of this Mexican version of a grilled cheese sandwich. This is a great lunch with a small salad. And we use the microwave for it.

4	9-inch (23 cm) flour tortillas	4
¾ cup	shredded Monterey Jack cheese	175 mL
¼ cup	Pickled Jalapeño Peppers (page 111), drained	50 mL
⅓ cup	low-fat plain yogurt	75 mL
⅓ cup	salsa (chapter 8) or commercial salsa	75 mL
1 tbsp	chopped fresh cilantro	15 mL

1. Place 1 tortilla on a microwavable plate lined with a double thickness of paper towel. Sprinkle with half of cheese; top with several slices of jalapeños. Press second tortilla over top. Microwave at Medium-High (70%) for 2 minutes or until tortillas are warm and cheese has melted. Cut into quarters with a sharp knife. Repeat with remaining tortillas.
2. Stir together yogurt, salsa and cilantro. Serve warm quesadillas with yogurt-salsa sauce.

Makes 8 pieces, or 2–3 servings.

Variations:

What you fold into the tortilla is up to your imagination, although the recipe above is fairly traditional. Other ideas are:
Vegetarian: Stir together 1 cup (250 mL) mashed kidney beans, 1 chopped green onion and 2 tbsp (25 mL) each salsa and plain yogurt. Spread evenly on tortilla and proceed as above.
French-Style: Spread goat cheese (chèvre) on tortilla. Sprinkle with several sliced Pickled Jalapeño Peppers (page 111), finely chopped red onion and chopped black olives. Proceed as above.
Open-Face Pizza: Spread homemade salsa on 1 tortilla, sprinkle with chopped onions, chopped sweet green peppers, sliced olives and shredded Cheddar cheese. Broil until cheese melts.
Pesto: Spread Presto Pesto (page 176) on tortilla. Sprinkle with chopped sweet yellow peppers, diced tomato and sliced green onions. Proceed as above.

Nippy Apple Cheddar Soup

Cheddar cheese and apples are as much of a go-together as any food we know. Here's a new twist: put them in a steaming bowl of soup with a hint of curry. Bound to hit the spot on a cold day.

1 tbsp	butter or margarine	15 mL
2	medium carrots, thinly sliced	2
1	onion, chopped	1
1	clove garlic, minced	1
2 tsp	each dry mustard and curry powder	10 mL
2 cups	chicken broth	500 mL
1 cup	apple juice or cider	250 mL
1 cup	Sweet and Chunky Apple Butter (page 79)	250 mL
	Salt and cayenne pepper	
1 cup	shredded old Cheddar cheese	250 mL

1. Melt butter in a medium saucepan over medium heat. Cook carrots, onion, garlic, mustard and curry powder for 5 minutes or until onion softens, stirring occasionally.
2. Add broth, apple juice and Apple Butter. Bring to a boil, reduce heat and boil gently for about 15 minutes or until vegetables are very tender; cool slightly. Process in a food processor or blender until smooth. Return to saucepan to reheat. Season to taste with salt and cayenne pepper. Spoon into bowls; sprinkle with cheese and serve.

Makes four 1-cup (250 mL) servings.

Country Corn Chowder

Taking the lid off Country Corn Relish helps you make this simple but delicious soup in a hurry. Its robust corn taste is ideal for casual meals.

3	slices bacon, diced	3
1	medium onion, chopped	1
2 tbsp	all-purpose flour	25 mL
2 cups	milk	500 mL
1	large potato, peeled and cubed	1
1	large carrot, thinly sliced	1
1 cup	Country Corn Relish (page 128)	250 mL
	Salt and pepper to taste	
2 tbsp	chopped fresh parsley	25 mL

1. Combine bacon and onion in a nonstick saucepan; cook on medium heat for 5 minutes or until onion is tender, stirring frequently. Stir in flour and cook for 30 seconds, stirring constantly. Gradually whisk in milk until smooth and thickened.
2. Add potato, carrot and Corn Relish. Cook on low heat for 20 minutes or until vegetables are tender, stirring often to prevent sticking.
3. Season to taste with salt and pepper; sprinkle each serving with parsley.

Makes four 1-cup (250 mL) servings.

Vinaigrettes and Dressings for Salads

Favourite Vinaigrette Dressing

Perfect vinaigrettes can be made from any combination of oil and vinegar. It's the proportions that are important. In addition to serving with assorted greens, drizzle over chilled cooked asparagus, carrots or long green beans.

¼ cup	flavoured oil (pages 190–193)	50 mL
2 tbsp	specialty vinegar (pages 183–189)	25 mL
2 tbsp	water	25 mL
1 tsp	Dijon mustard	5 mL
Pinch	each granulated sugar, salt and freshly ground pepper	Pinch

Combine all ingredients in a small container with a tight-fitting lid. Cover and shake well. Refrigerate until ready to use. Makes ½ cup (125 mL) vinaigrette.

Mango Chutney Vinaigrette

Mango Chutney's marvellous mango flavour makes a vinaigrette that is a natural with fruit salad and as a marinade for barbecued chicken, pork and ham.

½ cup	canola oil	125 mL
¼ cup	Mango Chutney (page 154)	50 mL
2 tbsp	rice vinegar	25 mL
2 tbsp	soy sauce	25 mL
2 tbsp	liquid honey	25 mL
1 tbsp	minced candied ginger	15 mL
	Grated rind and juice of 1 lime	

Whisk together oil, chutney, vinegar, soy sauce, honey, ginger, lime rind and lime juice. Refrigerate in a tightly sealed container. Makes about 1 cup (250 mL).

Honey Mustard Dressing

An excellent use for this dressing is to marinate chicken pieces before baking or barbecuing.

⅓ cup	olive oil	75 mL
¼ cup	red wine vinegar	50 mL
2 tbsp	Honey Mustard (page 163)	25 mL
1 tbsp	lemon juice	15 mL
	Freshly ground pepper	

Whisk together oil, vinegar, mustard and lemon juice. Add pepper to taste. Refrigerate in a tightly sealed container until ready to serve over crisp salad greens. Makes about ¾ cup (175 mL).

Creamy Mustard Dressing

Similar to the vinaigrette above, but a mayonnaise version for those times when you need a creamy dressing.

⅓ cup	Honey Mustard (page 163)	75 mL
⅓ cup	light mayonnaise	75 mL
2 tbsp	minced green onion	25 mL
1 tbsp	minced fresh parsley	15 mL
½ cup	canola oil	125 mL
¼ cup	cider vinegar	50 mL

Stir together mustard, mayonnaise, onion and parsley. Whisk in oil and vinegar. Refrigerate in a tightly sealed container until ready to use with cabbage, potato or pasta salads. Makes about 1¼ cups (300 mL).

Make-Ahead Salads

Spiced Orange Slices and Southwest Black Bean and Corn Salsa add instant flavour to salads that can be refrigerated and waiting until serving time.

Spiced Orange-Slice Salad

1½ cups	shredded romaine lettuce	375 mL
¼ cup	finely chopped red onion	50 mL
1 cup	frozen peas	250 mL
1	jar (half-pint/250 mL)	1
	Spiced Orange Slices (page 119)	
½ cup	sliced water chestnuts	125 mL
⅓ cup	light mayonnaise	75 mL
½–1 tsp	curry powder	2–5 mL
½ cup	shredded mozzarella cheese	125 mL
	Paprika to taste	

1. Arrange lettuce in a medium glass salad bowl. Top with onion and peas.
2. Drain liquid from oranges, reserving ⅓ cup (75 mL). Layer orange slices and water chestnuts over peas.
3. Combine reserved orange liquid with mayonnaise and curry powder; stir well. Spoon over salad; sprinkle with mozzarella and paprika. Cover tightly and refrigerate for several hours before serving.

Makes 6 servings.

Southwest Black Bean and Rice Salad

Combine 1 cup (250 mL) Southwest Black Bean and Corn Salsa, 1 cup (250 mL) cooked rice and ½ cup (125 mL) chopped sweet green pepper in a salad bowl. Whisk together 2 tbsp (25 mL) oil, 1 tbsp (15 mL) red wine vinegar and a pinch of salt; pour over salad. Chill until serving time. Makes 2 cups (500 mL).

Chicken with Apricot Sauce

Apricot Grand Marnier Conserve turns boneless chicken breasts into elegant fare.

2	boneless, skinless chicken breasts, halved	2
1 tsp	butter or margarine	5 mL
½ cup	Apricot Grand Marnier Conserve (page72)	125 mL
1	medium orange, peeled and chopped	1
3 tbsp	each brandy and chicken broth	45 mL
	Chopped fresh parsley	

1. Remove and discard fat from chicken. Melt butter in a large nonstick skillet over medium-high heat. Brown chicken quickly on each side.
2. Combine conserve, orange, brandy and chicken broth. Spoon over chicken, cover and cook on medium heat for 10 minutes or until chicken is no longer pink inside. Sprinkle with chopped parsley and serve.

Makes 4 servings.

Marmalade Squares

Any of our marmalades in chapter 3 can be successfully used in this fruitcake, although one with lots of citrus is probably best.

½ cup	each chopped dates, slivered dried apricots and diced candied pineapple	125 mL
½ cup	each dried cranberries, raisins and coarsely chopped walnuts	125 mL
1 cup	marmalade (chapter 3)	250 mL
½ cup	orange juice	125 mL
½ cup	mashed bananas	125 mL
1 cup	softened butter or margarine	250 mL
1 cup	lightly packed brown sugar	250 mL
3	eggs	3
2½ cups	all-purpose flour	625 mL
2 tsp	each baking powder and cinnamon	10 mL
½ tsp	ground ginger	2 mL
¼ tsp	salt	1 mL
	icing sugar	

1. Combine dates, apricots, pineapple, cranberries, raisins and walnuts in a large bowl. Stir in marmalade and orange juice. Cover and let stand overnight. Stir in banana.
2. Cream butter and sugar. Beat in eggs. Stir into fruit mixture.
3. Blend flour, baking powder, cinnamon, ginger and salt; stir into fruit mixture. Spoon into lightly greased 9 x 13-inch (3.5 L) baking dish. Bake in 325°F (160°C) oven for 45 minutes or until a cake tester inserted in centre comes out clean.
4. Cool completely; sprinkle with icing sugar.

Makes 16–24 pieces.

Apple Cranberry Mincemeat Squares

Just as tasty as mincemeat pie without the bother of making pastry. They freeze really well.

1¼ cups	all-purpose flour	300 mL
½ cup	cold butter or margarine, cut in pieces	125 mL
¼ cup	lightly packed brown sugar	50 mL
1 tsp	baking powder	5 mL
2	eggs	2
1 cup	granulated sugar	250 mL
⅓ cup	finely chopped walnuts	75 mL
2 tbsp	melted butter or margarine	25 mL
1 tbsp	milk	15 mL
1 tsp	vanilla extract	5 mL
1½ cups	Apple Cranberry Mincemeat (page 202)	375 mL

1. Place flour, butter, brown sugar and baking powder in a food processor. Process until butter is finely chopped. Separate 1 egg, reserving white. Add egg yolk to food processor and process until blended.
2. Pat mixture into bottom of an 8-inch (2 L) square baking dish. Bake in 350°F (180°C) oven for 10 minutes. Remove from oven.
3. Beat together egg and reserved egg white until frothy. Stir in sugar, nuts, melted butter, milk and vanilla.
4. Spread mincemeat over crust. Pour egg mixture over mincemeat. Bake for 25 minutes or until lightly browned. Cool and cut into squares.

Makes 16 squares.

Peach 'n' Jam Spice Upside-Down Cake

Old favourites never really go out of style, and this simple cake proves it. Use any fruit jam or conserve. Replace the peaches with apples for an equally delicious cake.

¼ cup	melted butter or margarine	50 mL
½ cup	Fruit jam or conserve	125 mL
¼ cup	firmly packed brown sugar	50 mL
1 tbsp	minced fresh peeled gingerroot	15 mL
12	slices peeled fresh peaches (3 peaches)	12
Cake:		
⅓ cup	melted butter or margarine	75 mL
¼ cup	molasses	50 mL
1	egg, beaten	1
¾ cup	firmly packed brown sugar	175 mL
2 cups	all-purpose flour	500 mL
1½ tsp	baking powder	7 mL
½ tsp	each: baking soda and ground cloves	2 mL
¼ tsp	ground allspice	1 mL
¾ cup	buttermilk	175 mL

1. Stir together ¼ cup (50 mL) melted butter, jam, brown sugar and gingerroot. Spoon into a lightly greased 9-inch (2.5 L) springform pan. Arrange peach slices over jam.

Cake:

1. Beat melted butter, molasses, egg and brown sugar in a medium bowl until light and creamy.
2. Blend flour, baking powder, baking soda, cloves and allspice in second bowl.
3. Add flour mixture alternately with buttermilk to creamed mixture, ending with flour, and beating well after each addition. Spoon batter over peaches. Bake in 350°F (180°C) oven for 40 minutes or until a tester inserted in centre comes out clean. Let cool for 10 minutes on a rack before inverting onto cake plate. Serve warm or at room temperature.

Makes 8 servings.

Sweet and Chunky Apple Butter Spice Cake

This easy one-bowl cake is full of fresh apple and spice flavours. Sweet and Chunky Apple Butter provides the moistness necessary to lower the fat yet maintain excellent texture and taste.

1 tbsp	butter or margarine	15 mL
1 cup	Sweet and Chunky Apple Butter (page 79)	250 mL
¾ cup	lightly packed brown sugar	175 mL
3 tbsp	canola oil	45 mL
½ cup	raisins	125 mL
½ cup	buttermilk	125 mL
1	egg, beaten	1
2 tsp	vanilla extract	10 mL
2¼ cups	all-purpose flour	550 mL
2 tsp	each ground cinnamon and ginger	10 mL
1 tsp	each baking soda, baking powder and ground allspice	5 mL
¼ tsp	salt	1 mL

1. Lightly spray a 9 x 13-inch (3.5 L) baking dish with vegetable spray.
2. Heat butter on medium-high heat in a small saucepan until it turns a nutty brown. Pour into a large mixing bowl. Add Apple Butter, sugar and oil. Stir until smooth.
3. Add raisins, buttermilk, egg and vanilla; mix well.
4. Combine flour, cinnamon, ginger, baking soda, baking powder, allspice and salt. Stir into wet ingredients just until combined. Pour into prepared pan.
5. Bake in 350°F (180°C) oven for 35 minutes or until cake tester inserted in centre comes out clean. Let cool for 10 minutes before cutting.

Makes 12 servings.

Microwave Jam Apple Crisp

Your favourite jam gives a wonderful flavour to this easy dessert. We especially like to use Blueberry Freezer Jam with Cointreau (page 32). Cooking the topping separately makes for a nicely browned topping without heating a conventional oven.

⅓ cup	all-purpose flour	75 mL
⅓ cup	brown sugar	75 mL
⅓ cup	quick-cooking rolled oats	75 mL
2 tbsp	butter or margarine	25 mL
½ cup	brown sugar	125 mL
3 tbsp	cornstarch	45 mL
½ cup	blueberry or other jam or marmalade	125 mL
4 cups	chopped, cored, peeled apples (4 to 5 apples)	1 L

1. Combine flour, ⅓ cup (75 mL) brown sugar, oats and butter in a bowl. Spread into an 8-inch (2 L) square microwavable baking dish; microwave on High (100%) for 4 minutes or until topping is browned and crumbly, stirring twice. Remove from dish and set aside.
2. Combine ½ cup (125 mL) brown sugar and cornstarch in a large microwavable container. Blend in jam and half the apples. Microwave, uncovered, on High (100%) for 3 minutes or until mixture is bubbly. Stir in remaining apples; spread in the baking dish.
3. Sprinkle reserved topping over apples and microwave, uncovered, on High for 5 minutes or until bubbly around edges. Cool and cut into pieces.

Makes 9 servings.

Variation:

Microwave Marmalade Rhubarb Crisp: Replace apples with chopped rhubarb and use English Seville, Five Citrus, or Ruby-Red Grapefruit Marmalade from chapter 3.

Tip: For a fast dessert, keep the topping made in step 1 on hand in a tightly closed jar. To make a single serving, slice 1 apple into a small microwavable bowl, mix with a bit of cinnamon and sugar and sprinkle with topping, Microwave at High (100%) for 2 minutes. Serve warm.

Lady Fingers
with Lemon Mousse

Our Microwave Lemon Curd folded into whipped cream and layered with lady fingers produces this delicate and frothy dessert.

1 cup	Microwave Lemon Curd (page 82)	250 mL
¾ cup	whipping cream	175 mL
2 tbsp	icing sugar	25 mL
1	pkg (85 g) soft lady fingers	1
2 tbsp	sweet sherry, divided	25 mL

1. Place curd in a bowl. In a separate bowl whip cream and icing sugar until soft peaks form. Fold into curd.
2. Cut each lady finger in half, lengthwise. Place half of lady fingers in a shallow bowl or pie plate and drizzle with 1 tbsp (15 mL) sherry. Top with half of curd mixture. Repeat layers with remaining lady fingers, sherry and curd mixture.
3. Cover and refrigerate for several hours.

Makes 4 to 6 servings.

That Special Gift

How often do you ask yourself, "What makes a good hostess gift other than the usual bottle of wine?" We suggest many of the sweet spreads and condiments in our book. Often, the most treasured gifts are something somebody took the time and trouble to make.

Naturally, gifts from the kitchen should suit the interests of the person receiving them. For instance, during the holiday season what could be more appropriate than something to help a busy host—a jar of mincemeat (page 202) or one of the special fruit curds (pages 72–82). Or consider one of our fruit liqueurs (pages 208–210). Dieters should appreciate a flavoured vinegar for flavour without calories (pages 183–189). Any host who enjoys breakfast will be delighted with one of your cranberry treasures—try Cranberry Orange Marmalade (page 62). Those with an active life are bound to appreciate a jar of Tapenade-Style Salsa (page 147) or Green and Red Pepper Relish (page 129) to serve at cocktail time.

Packaging the Treasures

Attractive packaging certainly enhances the thoughtfulness of your gift. *Creative* is the key word when it comes to decorating. Here are a few ideas:

- line wicker baskets with leftover scraps of fabric;
- tie raffia to bottle necks;
- use seasonal fabrics with pine cones and holly for Christmas giving;
- decorate lids with freehand paintings or "add ons" such as scrap fabric, wallpaper, burlap, beads or silk flowers;
- cover lids with an appropriate fabric tied on with matching ribbon;
- decorate jars with decals, paint or découpage;
- sew reusable drawstring gift bags from leftover fabric;
- spruce up wooden baskets, berry baskets or grocery-store and wine boxes with decorations and paint;
- personalize labels with handwriting or the graphics package on your home computer;
- wrap your gift in coloured cellophane or other interesting wrapping material;
- place your gift in a container that has a practical use such as a bamboo steamer, a colander, a sieve, a pottery mug or cup and saucer, or a barbecue fish basket.

These ideas, and the many more you will think of, will make your gifts doubly appreciated—and memorable. And if your gift is to another cook, enclose a copy of the recipe.

Index